Lecture Notes in Mathematics

Edited by J.-M. Morel, F. Takens and B. Teissier

Editorial Policy for Multi-Author Publications: Summer Schools / Intensive Courses

1. Lecture Notes aim to report new developments in all areas of mathematics and their applications – quickly, informally and at a high level. Mathematical texts analysing new developments in modelling and numerical simulation are welcome. Manuscripts should be reasonably self-contained and rounded off. Thus they may, and often will, present not only results of the author but also related work by other people. They should provide sufficient motivation, examples and applications. There should also be an introduction making the text comprehensible to a wider audience. This clearly distinguishes Lecture Notes from journal articles or technical reports which normally are very concise. Articles intended for a journal but too long to be accepted by most journals, usually do not have this "lecture notes" character.

2. In general SUMMER SCHOOLS and other similar INTENSIVE COURSES are held to present mathematical topics that are close to the frontiers of recent research to an audience at the beginning or intermediate graduate level, who may want to continue with this area of work, for a thesis or later. This makes demands on the didactic aspects of the presentation. Because the subjects of such schools are advanced, there often exists no textbook, and so ideally, the publication resulting from such a school could be a first approximation to such a textbook.
   Usually several authors are involved in the writing, so it is not always simple to obtain a unified approach to the presentation.
   For prospective publication in LNM, the resulting manuscript should not be just a collection of course notes, each of which has been developed by an individual author with little or no co-ordination with the others, and with little or no common concept. The subject matter should dictate the structure of the book, and the authorship of each part or chapter should take secondary importance. Of course the choice of authors is crucial to the quality of the material at the school and in the book, and the intention here is not to belittle their impact, but simply to say that the book should be planned to be written by these authors jointly, and not just assembled as a result of what these authors happen to submit.
   This represents considerable preparatory work (as it is imperative to ensure that the authors know these criteria before they invest work on a manuscript), and also considerable editing work afterwards, to get the book into final shape. Still it is the form that holds the most promise of a successful book that will be used by its intended audience, rather than yet another volume of proceedings for the library shelf.

3. Manuscripts should be submitted (preferably in duplicate) either to Springer's mathematics editorial in Heidelberg, or to one of the series editors (with a copy to Springer). Volume editors are expected to arrange for the refereeing, to the usual scientific standards, of the individual contributions. If the resulting reports can be forwarded to us (series editors or Springer) this is very helpful. If no reports are forwarded or if other questions remain unclear in respect of homogeneity etc, the series editors may wish to consult external referees for an overall evaluation of the volume. A final decision to publish can be made only on the basis of the complete manuscript, however a preliminary decision can be based on a pre-final or incomplete manuscript. The strict minimum amount of material that will be considered should include a detailed outline describing the planned contents of each chapter.
   Volume editors and authors should be aware that incomplete or insufficiently close to final manuscripts almost always result in longer evaluation times. They should also be aware that parallel submission of their manuscript to another publisher while under consideration for LNM will in general lead to immediate rejection.

Continued on inside back-cover

Lecture Notes in Mathematics 1862

Editors:
J.-M. Morel, Cachan
F. Takens, Groningen
B. Teissier, Paris

Bernard Helffer
Francis Nier

Hypoelliptic Estimates and Spectral Theory for Fokker-Planck Operators and Witten Laplacians

 Springer

Authors

Bernard Helffer
Laboratoire de Mathématiques
UMR CNRS 8628
Université Paris-Sud
Bâtiment 425
91404 Orsay
France
e-mail: bernard.helffer@math.u-psud.fr

Francis Nier
IRMAR UMR CNRS 6625
Université de Rennes 1
Campus de Beaulieu
35042 Rennes Cedex
France
e-mail: francis.nier@univ-rennes1.fr

Library of Congress Control Number: 2004117183

Mathematics Subject Classification (2000):
35H10, 35H20, 35P05, 35P15, 58J10, 58J50, 58K65, 81Q10, 81Q20, 82C05, 82C31, 82C40

ISSN 0075-8434
ISBN 3-540-24200-7 Springer Berlin Heidelberg New York
DOI: 10.1007/b104762

Springer is a part of Springer Science + Business Media
http://www.springeronline.com
© Springer-Verlag Berlin Heidelberg 2005
Printed in Germany

Typesetting: Camera-ready TeX output by the authors

41/3142/du - 543210 - Printed on acid-free paper

Foreword

This text is an expanded version of informal notes prepared by the first author for a minicourse of eight hours, reviewing the links between hypoelliptic techniques and the spectral theory of Schrödinger type operators. These lectures were given at Rennes for the workshop "Equations cinétiques, hypoellipticité et Laplacien de Witten" organized in February 2003 by the second author. Their content has been substantially completed after the workshop by the two authors with the aim of showing applications to the Fokker-Planck operator in continuation of the work by Hérau-Nier. Among other things it will be shown how the Witten Laplacian occurs as the natural elliptic model for the hypoelliptic drift diffusion operator involved in the kinetic Fokker-Planck equation. While presenting the analysis of these two operators and improving recent results, this book presents a review of known techniques in the following topics : hypoellipticity of polynomial of vector fields and its global counterpart, global Weyl-Hörmander pseudo-differential calculus, spectral theory of non self-adjoint operators, semi-classical analysis of Schrödinger type operators, Witten complexes and Morse inequalities.

The authors take the opportunity to thank J.-M. Bony, who permits them to reproduce its very recent unpublished results, and also M. Derridj, M. Hairer, F. Hérau, J. Johnsen, M. Klein, M. Ledoux, N. Lerner, J.M. Lion, H.M. Maire, O. Matte, J. Moeller, A. Morame, J. Nourrigat, C.A. Pillet, L. Rey-Bellet, D. Robert, J. Sjöstrand and C. Villani for former collaborations or discussions on the subjects treated in this text. The first author would like to thank the Mittag-Leffler institute and the Ludwig Maximilian Universität (Munich) where part of these notes were prepared and acknowledges the support of the European Union through the IHP network of the EU No HPRN-CT-2002-00277 and of the European Science foundation (programme SPECT). The second author visited the Mittag-Leffler institute in september 2002 and acknowledges the support of the french "ACI-jeunes chercheurs : Systèmes hors-équilibres quantiques et classiques", of the Région Bretagne, of Université de Rennes 1 and of Rennes-Métropole for the organization of the workshop "CinHypWit : Equations cinétiques, Hypoellipticité et Laplaciens de Witten" held in Rennes 24/02/03-28/02/03.

Contents

1

Introduction

This text presents applications and new issues for hypoelliptic techniques initially developed for the regularity analysis of partial differential operators. The main motivation comes from the theory of kinetic equation and statistical physics. We will focus on the Fokker-Planck (Kramers) operator:

$$K = v \cdot \partial_x - (\partial_x V(x)) \cdot \partial_v - \Delta_v + \frac{v^2}{4} - \frac{n}{2} = X_0 - \Delta_v + \frac{v^2}{4} - \frac{n}{2} , \quad (1.1)$$

and the Witten Laplacian

$$\Delta^{(0)}_{\Phi/2,h} := -h^2 \Delta + \frac{1}{4}|\nabla\Phi|^2 - \frac{h}{2}\Delta\Phi , \quad (1.2)$$

where

$$\Phi(x,v) = \frac{v^2}{2} + V(x)$$

is a classical hamiltonian on $\mathbb{R}^{2n}_{x,v}$ and

$$X_0 = v \cdot \partial_x - (\partial_x V(x)) \cdot \partial_v$$

is the corresponding hamiltonian vector field.
The aim of this text is threefold:

1. exhibit the strong relationship between these two operators,
2. review the known techniques initially devoted to the analysis of hypoelliptic differential operators and show how they can become extremely efficient in this new framework,
3. present, complete or simplify the existing recent results concerned with the two operators (1.1) and (1.2).

At the mathematical level the analysis of these two operators leads to explore or revisit various topics, namely: hypoellipticity of polynomials of vector fields and its global counterpart, global Weyl-Hörmander pseudo-differential calculus, spectral theory of non self-adjoint operators, semi-classical analysis of Schrödinger type operators, Witten complexes and Morse inequalities. The point of view chosen in this text is, instead of considering more complex physical models, to focus on these two operators and to push as far as possible the analysis. In doing so, new results are obtained and some new questions arise about the existing mathematical tools.

We will prove that $(e^{-tK})_{t\geq 0}$ and $(e^{-t\Delta_{\Phi/2}^{(0)}})_{t\geq 0}$ are well defined contraction semigroups on $L^2(\mathbb{R}^{2n}, dx\, dv)$ for any $V \in \mathcal{C}^\infty(\mathbb{R}_x^n)$. Meanwhile the Maxwellian

$$M(x,v) = \begin{cases} e^{-\frac{\Phi(x,v)}{2}} & \text{if } e^{-\frac{\Phi(x,v)}{2}} \in L^2(\mathbb{R}^{2n}) \\ 0 & \text{else}, \end{cases}$$

is the (unique up to normalization) equilibrium for K and $\Delta_{\Phi/2}^{(0)}$:

$$KM = \Delta_{\Phi/2}^{(0)}M = 0 .$$

Two questions arise from statistical physics or the theory of kinetic equations:

Question 1:
Is there an exponential return to the equilibrium ? By this, we mean the existence of $\tau > 0$ such that:

$$\left\| e^{-tP}u - c_u M \right\| \leq e^{-\tau t} \left\| u \right\| , \quad \forall u \in L^2(\mathbb{R}^{2n}) ,$$

where $P = K$ or $P = \Delta_{\Phi/2}^{(0)}$ and c_u (in the case $M \neq 0$) is the scalar product in $L^2(\mathbb{R}^{2n})$ of u and $M/\|M\|$.

Question 2:
Is it possible to get quantitative estimates of the rate τ ?

For $P = \Delta_{\Phi/2}^{(0)}$ which is essentially self-adjoint it is reduced to the estimate of its first nonzero eigenvalue. Several recent articles, like [DesVi], [EckPiRe-Be], [EckHai1], [EckHai2], [HerNi], [Re-BeTh1], [Re-BeTh2], [Re-BeTh3], [Ta1], [Ta2] and [Vil], analyzed this problem for operators similar to K, with various approaches going from pure probabilistic analysis to pure partial differential equation (PDE) techniques and to spectral theory. The point of view developed here is PDE oriented and will strongly use hypoelliptic techniques together with the the spectral theory for non self-adjoint operators.

Note that a related and preliminary result in this "spectral gap" approach concerns the compactness of the resolvent. One of the results which establish the strong relationship between K and $\Delta_{\Phi/2}^{(0)}$ says:

Theorem 1.1.
The implication

$$\left((1+K)^{-1} \ compact \right) \Rightarrow \left((1+\Delta_\Phi^{(0)})^{-1} \ compact \right) \tag{1.3}$$

holds under the only assumption $V \in \mathcal{C}^\infty(\mathbb{R}^n)$[1].

---

[1] Indeed the $\mathcal{C}^\infty$ regularity is not the crucial point here and the most important fact is that nothing is assumed about the behaviour at infinity.

In [HerNi] the reverse implication was proved for quite general elliptic potentials, satisfying for some $\mu \geq 1$,

$$|\partial_x^\alpha V(x)| \leq C_\alpha \langle x \rangle^{2\mu - |\alpha|} \text{ and } C^{-1} \langle x \rangle^{2\mu} \leq 1 + |V(x)| \leq C^\alpha \langle x \rangle^{2\mu} .$$

Among other things in the present text, we will explore as deeply as possible the validity of the following conjecture:

Conjecture 1.2.
The Fokker-Planck operator (1.1) has a compact resolvent if and only if the Witten Laplacian on 0-forms (1.2) has a compact resolvent.

Hypoelliptic techniques enter at this level twice:

1. in the proof of the equivalence when it is possible;
2. in order to get effective criteria for the compactness of $(1 + \Delta_{\Phi/2}^{(0)})^{-1}$.

In this direction, the present text provides a (non complete) review of various techniques due to Hörmander [Hor1], Kohn [Ko], Helffer-Mohamed [HelMo], Helffer-Nourrigat [HelNo1, HelNo2, HelNo3, HelNo4], while emphasizing new applications of rather old results devoted to subellipticity of systems by Maire [Mai1, Mai2], Trèves [Tr2] and Nourrigat [No1]. Among those works, one can distinguish at least two methods for the treatment of the hypoellipticity, one referred to as Kohn's method which is not optimal but flexible enough to permit several variants and another one which is based on the idea initiated by Rothschild-Stein [RoSt] and developed by Helffer-Nourrigat to approximate the operators by left invariant operators on nilpotent Lie groups.
By writing

$$\Delta_{\Phi/2}^{(0)} = \Delta_{V/2}^{(0)} \otimes \mathrm{Id}_v + \mathrm{Id}_x \otimes (-\Delta_v + \frac{v^2}{4} - \frac{n}{2})$$

and

$$\Delta_{V/2}^{(0)} = -\Delta_x + \frac{1}{4} |\nabla V|^2 - \frac{1}{2} \Delta V(x) ,$$

which can also be expressed in the form

$$\Delta_{V/2}^{(0)} = \sum_{j=1}^n L_j^* L_j = \sum_{j=1}^n X_j^2 + Y_j^2 + i\,[X_j, Y_j] ,$$

with $L_j = X_j + Y_j$, $X_j = \partial_{x_j}$, $Y_j = \frac{1}{2i} \partial_{x_j} V(x)$, the conditions on $V(x)$ which ensure the compactness of $(1 + \Delta_{\Phi/2}^{(0)})^{-1}$ can be analyzed very accurately with nilpotent techniques.
Although it is possible to write K as a non commutative polynomial of ∂_{x_j}, $\partial_{x_j} V(x)$, ∂_{v_j}, v_j, the relationship between K and $\Delta_{\Phi/2}^{(0)}$ is more clearly exhibited after writing

$$K = X_0 + b^* b = X_0 + \sum_{j=1}^n b_j^* b_j$$

which looks like a "type 2 Hörmander's operators", $X_0 + \sum_{j=1}^n Y_j^2$ if one replaces the vector field Y_j by the annihilation operators

$$b_j = \partial_{v_j} + \frac{v_j}{2}\ , \text{ for } j = 1, \ldots, n\ ,$$

associated with the harmonic oscillator hamiltonian

$$b^* b = -\Delta_v + \frac{v^2}{4} - \frac{n}{2}\ .$$

We will follow and improve the variant of the Kohn's method used by Hérau-Nier in [HerNi] which was partly inspired by former works of Eckmann-Pillet-Rey-Bellet [EckPiRe-Be], Eckmann-Hairer [EckHai1]. Precisely our results will require one of the two following assumptions after setting

$$h(x) = \sqrt{1 + |\nabla V(x)|^2}\ .$$

Assumption 1.3.
The potential $V(x)$ belongs to $C^\infty(\mathbb{R}^n)$ and satisfies:

$$\forall \alpha \in \mathbb{N}^n,\ |\alpha| \geq 1, \exists C_\alpha\ s.t.\ \forall x \in \mathbb{R}^n\ ,\ |\partial_x^\alpha V(x)| \leq C_\alpha h(x)\ , \qquad (1.4)$$

$$\exists M\ ,\ C \geq 1\ ,\ s.t\ \forall x \in \mathbb{R}^n\ ,\ h(x) \leq C \langle x \rangle^M\ , \qquad (1.5)$$

and the coercivity condition

$$\exists M\ ,\ C \geq 1\ ,\ s.t.\ \forall x \in \mathbb{R}^n\ ,\ C^{-1} \langle x \rangle^{1/M} \leq h(x)\ . \qquad (1.6)$$

Assumption 1.4.
The potential $V(x)$ belongs to $C^\infty(\mathbb{R}^n)$ and satisfies (1.4) (1.5) with the coercivity condition (1.6) replaced by the existence of $\rho_0 > 0$ and $C > 0$ such that:

$$\forall x \in \mathbb{R}^n,\ |\nabla h(x)| \leq C\, h(x) \langle x \rangle^{-\rho_0}\ . \qquad (1.7)$$

Theorem 1.5.
If the potential $V \in C^\infty(\mathbb{R}^n)$ verifies Assumption 1.3 or Assumption 1.4, then there exists a constant $C > 0$ such that

$$\forall u \in \mathcal{S}(\mathbb{R}^{2n}),\ \left\| \Lambda^{1/4} u \right\|^2 \leq C \left(\|Ku\|^2 + \|u\|^2 \right)\ , \qquad (1.8)$$

with $\Lambda^2 = (1 + \Delta_{\Phi/2}^{(0)})$

Corollary 1.6.
If the potential $V \in C^\infty(\mathbb{R}^n)$ satisfies Assumption 1.3 then the operator K has a compact resolvent.
If the potential $V \in C^\infty(\mathbb{R}^n)$ satisfies Assumption 1.4, then K has a compact resolvent if (and only if) the Witten Laplacian $\Delta_{V/2}^{(0)}$ has a compact resolvent.

After the proof of these results, we show by analyzing the example of a quadratic potential V that the exponent $1/4$ is not optimal. We also address the question whether nilpotent algebra method can be applied directly to the operator K and explain why a naive application of Helffer-Nourrigat results in [HelNo3] does not work. We emphasize that the hypoelliptic estimate (1.8) is not only used for the question of the compactness of $(1 + K)^{-1}$. Indeed a variant of it permits to give a meaning to the contour integral

$$e^{-tK} = \frac{1}{2i\pi} \int_{\partial S_K} e^{-tz}(z - K)^{-1} \, dz \,,$$

for $t > 0$, although we cannot say more on the numerical range of K, than

$$\{\langle u, Ku \rangle, \ u \in D(K)\} \subset \{z \in \mathbb{C}, \operatorname{Re} z \geq 0\} \,.$$

This last point is crucial in the quantitative analysis of the rate of return to the equilibrium.

We will not reproduce the complete quantitative analysis of [HerNi] which provides upper and lower bounds of the rate of return to the equilibrium for

$$K_{\gamma_0,m,\beta} = v \cdot \partial_x - \frac{1}{m}(\partial_x V(x)) \cdot \partial_v - \frac{\gamma_0}{m\beta}\left(\partial_v - \frac{m\beta}{2}v\right) \cdot \left(\partial_v + \frac{m\beta}{2}v\right)$$

in terms of the friction coefficient γ_0, the particle mass m and the inverse temperature β. These bounds are expressed, up to some explicit algebraic factor in (γ_0, m, β), in terms of the first non zero eigenvalue of the semiclassical Witten Laplacian

$$\Delta^{(0)}_{V/2,h} = -h^2 \Delta_x + \frac{1}{4}|\nabla V(x)|^2 - \frac{h}{2}\Delta V(x) \quad \text{with } h = \beta^{-1} \,.$$

The latter part of this text gives an account of the semiclassical analysis of this Witten Laplacian. We will recall the relationship with Morse inequalities according to Witten [Wi], after introducing the whole Witten complex and the corresponding deformed Hodge Laplacians $\Delta^{(p)}_{f,h}$ on all p-forms. After recalling some basic tools in semiclassical analysis, we recall the more accurate results of Helffer-Sjöstrand [HelSj1, HelSj4] stating that the $\mathcal{O}(h^{3/2})$ eigenvalues of these Witten Laplacians are actually $\mathcal{O}(e^{-\frac{S}{h}})$ and that the restriction of the Witten complex, to suitable finite dimensional spectral spaces, leads by a limiting procedure to the orientation complex which was introduced in topology. Finally, we will discuss and propose some improvements about the accurate asymptotics of those exponentially small eigenvalues given, by Bovier-Eckhoff-Gayrard-Klein in [BovGayKl], [BovEckGayKl1] and [BovEckGayKl2]. This last result will at the end be combined with the comparison inequalities of [HerNi] for the rates of trend to the equilibrium between $K_{\gamma_0,m,\beta}$ and $\Delta^{(0)}_{V/2,h}$, $(h = \beta^{-1})$.

Here is an example of quantitative results which can be obtained.

Proposition 1.7.
Assume that the potential V is a C^∞ Morse function with

- *two local minima $U_1^{(0)}$ and $U_2^{(0)}$, such that $V(U_1^{(0)}) < V(U_2^{(0)})$,*
- *one critical point with index 1 $U^{(1)}$,*
- *$V(x) = |x|^2$ for $|x| \geq C$.*

Then for fixed any fixed $\gamma_0 > 0$ and $m > 0$, the rate $\tau(\gamma_0, m, \beta)$ satisfies

$$\liminf_{\beta \to \infty} e^{\beta\left(V(U^{(1)}) - V(U_2^{(0)})\right)} \tau(\gamma_0, m, \beta) > 0 \,,$$

$$and \ \limsup_{\beta \to \infty} \frac{e^{\frac{\beta}{2}\left(V(U^{(1)}) - V(U_2^{(0)})\right)} \tau(\gamma_0, m, \beta)}{\beta \log \beta} < +\infty \,.$$

At the level of the methods, there is no strict separation between the qualitative and the quantitative analysis. This is especially true for the maximal estimates obtained for operators on nilpotent Lie algebra: the existence of uniform estimates can indeed lead by a kind of addition of variable procedure standard in physics to semi-classical estimates.

In order to help the reader who is not necessarily specialist in all the techniques, we now give a rather precise description of the contents of the book, chapter by chapter. We mention in particular the possibilities for the reader to omit some part at the first reading.

- In Chapter 2, we present the Hörmander condition for a family of vector fields and the proof given by J. Kohn of the subellipticity of the Hörmander's operators $\sum X_j^2$ and $X_0 + \sum X_j^2$. Although it is a rather standard material, we thought that it was useful to give the details because many other proofs will be modelled on this first one. The use of the pseudo-differential theory is minimal in this chapter, and appears essentially only for operators of the form $\Lambda^s := \phi(x)(1 - \Delta)^s\chi(x)$, composed with partial differential operators. We give all the details for the brackets arguments but do not recall how the hypoellipticity can be derived from these subelliptic estimates.
- In Chapter 3, we recall some basic criteria for the compactness of the resolvent of the Schrödinger operator following a paper of Helffer-Mohamed. Again, this is rather standard material but we show how to use the Kohn's argument in the context of global problems. The bracket's technique is used here in order to prove that the form domain of the Schrödinger operator is compactly embedded in L^2. This is simply obtained by showing the continuous imbedding of the form domain in a weighted L^2 space. We have not resisted to the pleasure to present the connected problem of the magnetic bottles.

- In Chapter 4, we recall some elements of the Weyl-Hörmander calculus. The main aim is to construct the analog of the Λ^s appearing in Kohn's proof in a very large context. Because we wanted here to extend as much as possible the previous work of Hérau-Nier in [HerNi], we were naturally led to introduce a rather general class of pseudo-differential operators adapted to this problem. The reader can at the first reading omit this chapter and just take the main result as a fact. The existence of this family $(\Lambda^s)_{s\in\mathbb{R}}$ of pseudo-differential operators when Λ is a globally elliptic or globally quasi-elliptic operator (whose simplest example is the square root of the harmonic oscillator) is rather old (See for example the work by D. Robert in the seventies). Here the Beals criterion in the framework of Weyl-Hörmander calculus allows to consider once and for all possibly degenerate cases. We close the discussion by presenting new results of J.-M. Bony about the geodesic temperance.

- Chapter 5 is the first key chapter. We first show that our Fokker-Planck operators are maximally accretive by extending a self-adjointness criterion of Simader. This result seems to be new. We then analyze various properties of the Fokker-Planck operator. The main point is the analysis of the compactness of the resolvent. Developing an approach initiated by Hérau-Nier and implementing the family Λ^s analyzed in the previous chapter, the proof is a tricky mixture between Kohn's proof of subellipticity, Helffer-Mohamed's proof for the compactness of the resolvent of the Schrödinger operator and of the algebraic structure of the Fokker-Planck operator. The link with a Witten Laplacian is emphasized and this leads to propose a natural necessary and sufficient condition for the compactness of the resolvent of the Fokker-Planck operator which is partially left open. This disproves also that only an Hörmander's type global condition is sufficient. We also analyze carefully the so called quadratic model, recalling on one hand the explicit computations presented in the book by Risken and showing on the other hand how "microlocal analysis" can be used for improving Kohn's type estimates.

- Chapter 6 shows how the previous hypoelliptic estimates permit to control the decay of the semi-group attached to the Fokker-Planck operator. The reader will find here the main motivation coming from the Kinetic theory. Again, we meet, when trying to be more quantitative, the question of estimating carefully the behavior of the lowest non zero eigenvalue of a canonical Witten Laplacian.

- Chapter 7 is devoted to a short description (without proofs) of the characterization of the hypoellipticity for homogeneous operators on nilpotent groups. The main result is a conjecture of Rockland which was proved in the late 70's by Helffer-Nourrigat. The reason for including this presentation in the book is two fold. First the hypoellipticity plays an important role in the analysis of the Fokker-Planck operator and the Witten Laplacian with degenerate ellipticity. Secondly, we consider maximal estimates and the proof of Helffer-Nourrigat was actually establishing as a technical tool a lot of spectral estimates for operators with polynomial coefficients.

- Chapter 8 develops the relationship between the nilpotent analysis and the more general analysis of maximal hypoellipticity of polynomial of vector fields. The breakthrough was the paper by Rothschild-Stein which opened the possibility to establish and prove good criteria of maximal hypoellipticity. We very briefly present some ideas of the results obtained in this spirit by Helffer-Nourrigat and Nourrigat during the eightie's.

- Chapter 9 is a first try to apply nilpotent techniques directly to the Fokker-Planck operator. We present the main difficulties and discuss various possible approaches. As an application of these ideas we obtain a first result containing the quadratic Fokker-Planck model, which is far from proving the general conjecture, but leads to optimal estimates.

- Chapter 10 presents how the nilpotent techniques work for particular systems. Instead of looking at the Witten Laplacian, it is better to look at the system corresponding to the first distorted differential of the Witten complex. The analysis of the microlocal maximal hypoellipticity or of the microlocal subellipticity of these systems of complex vector fields, which was done in the eighties mainly motivated by the $\bar{\partial}_b$-problem in complex analysis, gives as byproducts new results for the compactness of the resolvent and for the semi-classical regime. Following a former lecture note of Nourrigat, our presentation (without proof) of the basic results in microlocal analysis can be understood independently of the nilpotent language.

- Chapter 11 is continuing the investigation of the Witten Laplacian on $\mathbb{R}^n$. After recalling its general properties and its relationship with statistical mechanics (this point is detailed in Chapter 12), we present recent criteria for the compactness of its resolvent obtained by the authors and discuss many examples. New results are presented in connection with the subellipticity of some tangential system of vector fields.

- With Chapter 12, we start the presentation of the semi-classical analysis. The chapter is mainly devoted to the analysis of the so called harmonic approximation and we give a flavour of what is going on for large dimension systems which appear naturally in statistical physics.

- Chapter 13 enters more deeply in the analysis of the tunneling effect. Because there are already pedagogical books on the subject, we choose to select some of the important ideas and limit ourselves to the treatment of the first model of the theory: the double well problem.

- Chapter 14 starts the analysis of the Witten Laplacian in the semi-classical regime. We recall how E. Witten uses the harmonic approximation technique for suitable Laplacians on p-forms attached to a distorted complex of the de Rham complex in order to give an analytic proof of the Morse inequalities.

- Chapter 15 is again a key chapter. We now would like to analyze exponentially small effects. We recall (in a sometimes sketchy way) the main steps of the so called Witten-Helffer-Sjöstrand's proof that the Betti numbers are also the cohomology numbers of the orientation complex.

- Chapter 16 explores how this approach permits to understand and partially recover some recent results by Bovier-Gayrard-Klein. We also present the recent results obtained in collaboration with M. Klein. We close the chapter by an application to the splitting for the Witten Laplacian on functions.
- Chapter 17 is devoted to the presentation of the result obtained by Hérau-Nier for the rate of decay for the semi-group associated to the Fokker-Planck operators which was one of the main motivations of the whole study.
- The last chapter gives additional information on quite recent results obtained or announced in the last year.

2

Kohn's Proof of the Hypoellipticity of the Hörmander Operators

2.1 Vector Fields and Hörmander Condition

We consider p C^∞ real vector fields $(X_1, \cdots, X_p)$ in a open set Ω of $\mathbb{R}^n$. If X and Y are two vector fields, the bracket of X and Y, denoted by $[X, Y]$ or $(\operatorname{ad} X) Y$, is defined by

$$[X, Y]f = X(Yf) - Y(Xf).$$

We note that $[X, Y]$ is a new vector field. We are interested in the case when the Hörmander condition [Hor1] is satisfied.

Definition 2.1. Hörmander Condition
We say that the Hörmander condition is satisfied at x_0, if there exists $r(x_0) \geq 1$ such that the vector space generated by the iterated brackets $(\operatorname{ad} X)^\alpha X_k$ at x_0 with $|\alpha| \leq r(x_0) - 1$ is $\mathbb{R}^n$.

When $r(x_0) = 1$, we say that the system is elliptic and this imposes of course $p \geq n$. Let us give typical examples.

Heisenberg algebra:

$$n = 3\,,\ p = 2\,,\ r = 2\,,$$
$$X_1 = \partial_x, X_2 = x\partial_z + \partial_y\,,$$
$$[X_1\,,\ X_2] = \partial_z\,. \tag{2.1}$$

Grushin's operator:

$$n = 2\,,\ r = 2\,,$$
$$X_1 = \partial_x\,,\ X_2 = x\partial_y\,,$$
$$[X_1, X_2] = \partial_y\,.$$

Nilpotent group $\mathcal{G}_4$:

$$n = 4 \, , \; r = 3 \, ,$$
$$X_1 = \partial_x \, , \; X_2 = \tfrac{1}{2}x^2\partial_t + x\partial_z + \partial_y \, ,$$
$$[X_1, X_2] = x\partial_t + \partial_z \, ,$$
$$[X_1, [X_1, X_2]] = \partial_t \, .$$

We say that the vector fields X_j satisfy the Hörmander condition of rank r in an open set Ω if $r_{min}(x_0) \leq r$, for all $x \in \Omega$.

2.2 Main Results in Hypoellipticity

We first start by recalling the basic definition of hypoellipticity introduced by L. Schwartz:

Definition 2.2.
A differential operator with C^∞ coefficients in an open set Ω is hypoelliptic in Ω, if, for any $\omega \subset \Omega$, any $u \in \mathcal{D}'(\Omega)$, such that $Pu \in C^\infty(\omega)$, belongs to $C^\infty(\omega)$.

This terminology was motivated by the fact that the elliptic operators have this property and that in the fifties a very natural question was to give a characterization of the hypoellipticity for the operators with constant coefficients. Once this was settled, the second challenge was to understand the hypoellipticity of non necessarily elliptic operators with variable coefficients and the next theorem was probably one of the first general results in this direction.

Theorem 2.3.
If the vector fields (X_j) $(j = 1, \ldots, p)$ satisfy the Hörmander condition for some r in Ω, then the operator:

$$L = \sum_{j=1}^{p} X_j^2 \, , \tag{2.2}$$

*which will be called "**type 1 Hörmander's operator**", is hypoelliptic in Ω.*

This result is due to L. Hörmander [Hor1].

Remark 2.4.
The Hörmander condition is a necessary condition for getting hypoellipticity in the case when the X_j's have analytic coefficients. The proof (due to Derridj [Der]) is based on Nagano's Theorem. In the C^∞ case, the hypoelliptic operator $-\frac{d^2}{dx^2} - \exp-\frac{1}{x^2}\frac{d^2}{dy^2}$ in $\mathbb{R}^2$ shows that the Hörmander condition (which is not satisfied when $x = 0$) is not in general necessary.

Theorem 2.5.
*If the Hörmander condition is satisfied for some r in Ω, then, for any compact
subset $K \subset \Omega$, there exists C_K such that*

$$||u||_{\frac{1}{r}}^2 \le C \left(\sum_j ||X_j u||_0^2 + ||u||_0^2 \right) , \quad \forall u \in C_0^\infty(\Omega) , \quad with \ \operatorname{supp} u \subset K ,$$

$$(2.3)$$

where $||u||_s$ is the Sobolev norm corresponding to H^s.

Remark 2.6.
One can actually show (see for example [BoCaNo]) that the validity of the
inequality (2.3) (also called subelliptic estimates) implies the Hörmander condition of rank r.

Remark 2.7.
Except for the case of operators with constant coefficients where the construction of a suitable fundamental solution can lead directly to the proof of the
hypoellipticity, one is usually obliged[1] to get the C^∞ regularity by showing
that $u \in H_{loc}^s(\omega)$ for any s. It is indeed standard that $C^\infty(\omega) = \cap_{s \in \mathbb{R}} H_{loc}^s(\omega)$.
The proof of H^s-regularity is obtained through the proof of a priori estimates
in Sobolev spaces fo regular functions. The subelliptic estimate above is the
starting point for getting a complete family of inequalities of the type

$$||u||_{s+\frac{1}{r}}^2 \le C \left(\sum_j ||X_j u||_s^2 + ||u||_s^2 \right) , \quad \forall u \in C_0^\infty(\Omega) , \quad with \ \operatorname{supp} u \subset K ,$$

$$(2.4)$$

for any $s > 0$.

Remark 2.8.
There exists a microlocal version of this inequality which is due to Bolley-
Camus-Nourrigat [BoCaNo]. We will give precise definitions later in Chap-
ter 10.

Remark 2.9.
Note that it is immediate to see that (2.3) implies the same inequality with
X_j replaced by $X_j + c_j$ where the c_j's are C^∞ functions.

We also have to consider "**type 2 Hörmander's operators**", corresponding to:

$$L = \sum_{j=1}^p X_j^2 + X_0 ,$$

$$(2.5)$$

where the vector fields $(X_0, X_1, ..., X_p)$ satisfy the Hörmander condition. The
simplest example is the heat equation:

---
[1] Unless one constructs Parametrices

$$-\Delta_{x_1,\ldots,x_{n-1}} + \partial_{x_n}$$

A more typical case is the Kolmogorov operator [Kol]:

$$-\partial_x^2 + (x\partial_z + \partial_y) \,,$$

which was the motivating example for the analysis of Hörmander [Hor1]. As seen in the introduction, the motivating models like (1.1) are actually of this type.

2.3 Kohn's Proof

This section will be devoted to Kohn's proof [Ko] of some subelliptic estimates. It is simpler than the initial proof of Hörmander [Hor1] and permits other extensions. As a corollary, but this needs extrawork the existence of such inequalities imply the hypoellipticity of the corresponding Hörmander operator. These estimates are not optimal, in the sense that $\frac{1}{r}$ in the left hand side of (2.3) is replaced by the weaker 2^{-r}. Finally, let us emphasize that we are more interested in describing how the proof is going than in the result of hypoellipticity which is nowadays rather standard.

We consider the operator (2.5). The starting point is to get

$$\sum_{j=1}^{p} ||X_j u||^2 \le C \left(|\operatorname{Re} \langle Lu \mid u \rangle| + ||u||_0^2 \right) \,, \ \forall u \in C_0^\infty(V) \,, \tag{2.6}$$

where V is an open set.

This inequality is immediate by integration by parts if one observes that

$$X_j^* = -X_j + c_j \,, \tag{2.7}$$

for a C^∞ function c_j and that:

$$2|\operatorname{Re}\langle X_0 u \mid u \rangle| = |\langle c_0 u \mid u \rangle| \le C||u||_0^2 \,.$$

A Cauchy-Schwarz argument permits then to conclude.

We observe that this inequality of course implies:

$$\sum_{j=1}^{p} ||X_j u||^2 \le C \left(|\langle Lu \mid u \rangle| + ||u||_0^2 \right) \,, \tag{2.8}$$

and

$$\sum_{j=1}^{p} ||X_j u||^2 \le C \left(||Lu||_0^2 + ||u||_0^2 \right) \,. \tag{2.9}$$

Note that some information is lost in (2.9) in comparison with (2.6).

There is a general proof establishing that the subelliptic estimates (2.3) (or some weaker subelliptic estimate) joint with this inequality gives the hypoellipticity. The critical point in this part of the proof that we omit is the control of commutators of L with pseudo-differential operators.

In the case $X_0 = 0$ (the case $X_0 \neq 0$ requires more attention), the two inequalities (2.3) and (2.9) yield for some $\epsilon > 0$:

$$||u||_\epsilon^2 \leq C \left(||Lu||_0^2 + ||u||_0^2 \right) , \qquad (2.10)$$

but this inequality alone is not enough for proving hypoellipticity.

We now concentrate on the proof of the above subelliptic estimates, written in the form:

$$||u||_\epsilon^2 \leq C \left(||Lu||_0^2 + ||u||_0^2 \right) , \ \forall u \in C_0^\infty(V) , \qquad (2.11)$$

where $\epsilon > 0$ and V is a fixed open set containing the point in the neighborhood of which we want to show the hypoellipticity.

Although the general theory of pseudo-differential operators is not completely necessary, let us briefly recall that the pseudo-differential operators are operators which are defined by $u \mapsto \mathrm{Op}\,(a)u$ with:

$$\mathrm{Op}\,(a)u(x) := \frac{1}{(2\pi)^n} \int_{\mathbb{R}^n} \exp ix \cdot \xi \ a(x,\xi) \ \hat{u}(\xi) \, d\xi . \qquad (2.12)$$

Here $\mathbb{R}^{2n} \ni (x,\xi) \mapsto a(x,\xi)$ is a C^∞-symbol which admits as $|\xi| \to +\infty$ an expansion in homogeneous terms with respect to the ξ variables:

$$a(x,\xi) \sim \sum_{j \geq 0} a_{m-j}(x,\xi) ,$$

with

$$a_\ell(x,\lambda\xi) = \lambda^\ell a_\ell(x,\xi) , \forall \lambda > 0 , \forall \xi \in \mathbb{R}^n \setminus \{0\} .$$

The real number m is called the degree of the symbol (or of the corresponding pseudo-differential operator). Actually we only need here the composition of operators which are the multiplications by C^∞ functions, the differentiations and the family of convolution operators Λ^s, $s \in \mathbb{R}$, where Λ^s corresponds to the symbol $\langle\xi\rangle^s = (1 + |\xi|^2)^{\frac{s}{2}}$. When $s = 1$, we simply write Λ.

The important point is that the composition of two pseudo-differential operator of order m_1 and m_2 is a pseudo-differential operator of order $(m_1 + m_2)$ whose principal symbol is the product of the two principal symbols. Pseudo-differential operators of order 0 form an algebra of bounded operators in $\mathcal{L}(L^2(\mathbb{R}^n))$.

Now let $\mathcal{P}$ be the set of all pseudo-differential operators of order 0 such that if $P \in \mathcal{P}$, then there exists $\epsilon > 0$ and $C > 0$ such that:

$$||Pu||_\epsilon^2 \leq C \left(||Lu||_0^2 + ||u||_0^2 \right) , \ \forall u \in C_0^\infty(V) . \qquad (2.13)$$

This set satisfies the following properties:

Property 2.10.
(P1) $\mathcal{P}$ is a left and right ideal in the set of all pseudo-differential operators of order 0.

Property 2.11.
(P2) $\mathcal{P}$ is stable by taking the adjoint.

Property 2.12.
(P3) $X_j \Lambda^{-1} \in \mathcal{P}$ for $j = 0, \ldots, p$.

Property 2.13.
(P4) If $P \in \mathcal{P}$ then $[X_j, P] \in \mathcal{P}$ for $j = 0, \ldots, p$.

Let us first observe that one can prove inductively starting from $(P4)$ that $X_{i_1 \cdots i_p} \Lambda^{-1} \in \mathcal{P}$, with $X_{i_1 \cdots i_p} = [X_{i_1}, [X_{i_2}, \ldots [X_{i_{p-1}}, X_{i_p}] \ldots]]$. Let us for example show that $[X_j, X_k] \Lambda^{-1}$ belongs to $\mathcal{P}$. We know that $[X_j, X_k \Lambda^{-1}]$ has the property. But

$$[X_j, X_k \Lambda^{-1}] = [X_j, X_k] \Lambda^{-1} + X_k [X_j, \Lambda^{-1}].$$

Now

$$X_k [X_j, \Lambda^{-1}] = X_k \Lambda^{-1} (\Lambda[X_j, \Lambda^{-1}]),$$

and the operator $\Lambda[X_j, \Lambda^{-1}]$ is a pseudo-differential operator of order 0. Using $(P1)$, we get that $X_k \Lambda^{-1} (\Lambda[X_j, \Lambda^{-1}])$ belongs to $\mathcal{P}$.
Hence, using Hörmander condition of rank r, we deduce that $\mathcal{P}$ contains any pseudo-differential operator of order 0. It remains to prove the properties (P_j).

Proof of $(P1)$
It is a left ideal because pseudo-differential operators of order 0 are bounded in any Sobolev space. It is a right ideal as well owing to the property (P_2).

Proof of (P_2)
It is enough to observe that if P is a pseudo-differential operator of order 0, then

$$||\Lambda^\epsilon P^* u||^2 = \langle P\Lambda^{2\epsilon} P^* u \mid u \rangle = ||\Lambda^\epsilon P u||^2 + \langle (P\Lambda^{2\epsilon} P^* - P^* \Lambda^{2\epsilon} P) u \mid u \rangle.$$

We conclude by noticing that $(P\Lambda^{2\epsilon} P^* - P^* \Lambda^{2\epsilon} P)$ is a pseudo-differential operator of order $-1 + 2\epsilon \leq 0$ if $\epsilon \leq \frac{1}{2}$.

Proof of $(P3)$

For $j > 1$, we have

$$||\Lambda^{-1} X_j u||_\epsilon^2 \leq C(||X_j u||^2 + ||u||^2),$$

if $\epsilon \leq 1$. One can then conclude that $\Lambda^{-1} X_j \in \mathcal{P}$. Now we observe that $X_j \Lambda^{-1} = -(\Lambda^{-1} X_j)^* + c_j \Lambda^{-1}$. We then use $(P2)$ and we obtain $X_j \Lambda^{-1} \in \mathcal{P}$.

The treatment of the case of X_0 is a little more delicate.
We start from:

$$||\Lambda^{-1}X_0u||_{\frac{1}{2}}^2 = \langle X_0u \mid Tu \rangle , \qquad (2.14)$$

where T is a pseudo-differential operator of order 0.
Then we write $X_0u = Lu - \sum_j X_j^2u$, which leads to the estimate:

$$||\Lambda^{-1}X_0u||_{\frac{1}{2}}^2 \leq |\langle Lu \mid Tu \rangle| + \sum_{j>0} |\langle X_j^2u \mid Tu \rangle| . \qquad (2.15)$$

The first term of the right hand side is controlled. Let us show how the second one is treated. We have:

$$|\langle X_j^2u \mid Tu \rangle| = |\langle X_ju \mid X_j^*Tu \rangle| \leq C||X_ju||(||X_jTu|| + ||u||) .$$

Then we observe that

$$||X_jTu|| \leq (||X_ju|| + ||[X_j,T]u||) \leq C\left(||X_ju|| + ||u||\right) .$$

So we have shown that $\Lambda^{-1}X_0$ belongs to $\mathcal{P}$ with $\epsilon = \frac{1}{2}$. Taking the adjoint and observing that a pseudo-differential operator of strictly negative order belongs to $\mathcal{P}$ we get the result.

Proof of $(P4)$
Let us start from a P such that (2.11) holds for some $\epsilon > 0$.

The case $j > 0$.
Now consider:

$$||[X_j,P]u||_\delta^2 = \langle [X_j,P]u \mid \Lambda^{2\delta}[X_j,P]u \rangle$$
$$= \langle X_jPu \mid T^{2\delta}u \rangle - \langle PX_ju \mid T^{2\delta}u \rangle ,$$

where $T^{2\delta}$ is a pseudo-differental operator of order 2δ. It then follows that:

$$|\langle PX_ju \mid T^{2\delta}u \rangle| \leq |\langle X_ju \mid P^*T^{2\delta}u \rangle|$$
$$\leq ||X_ju||^2 + ||P^*T^{2\delta}u||^2$$
$$\leq ||X_ju||^2 + ||T^{2\delta}Pu||^2 + C||u||_{2\delta-1}^2 .$$

Similarly, with (2.7):

$$|\langle X_jPu \mid T^{2\delta}u \rangle| \leq |\langle Pu \mid X_jT^{2\delta}u \rangle| + C||Pu||_{2\delta}||u||_0$$
$$\leq C||Pu||_{2\delta}||X_ju||_0 + |\langle Pu \mid [X_j,T_{2\delta}]u \rangle| + C||Pu||_{2\delta}||u||_0 .$$

It remains to observe that:

$$|\langle Pu \mid [X_j,T_{2\delta}]u \rangle| \leq C||Pu||_{2\delta}||u|| .$$

Hence, the j's, $j > 0$, are done by choosing $\delta \leq \min(\frac{1}{2}, \frac{\epsilon}{2})$.

The case $j = 0$.
It is a little more delicate. We write, with (2.7) and $\delta \leq 1/2$,

$$
\begin{aligned}
|\langle X_0 Pu \mid T^{2\delta} u \rangle| &\leq |\langle Pu \mid T^{2\delta} X_0 u \rangle| \\
&\quad + C\|Pu\|_{2\delta}^2 + C\|u\|_0^2 \\
&\leq |\langle Pu \mid T^{2\delta} Lu \rangle| + \sum_{j>0} |\langle Pu \mid T^{2\delta} X_j^2 u \rangle| \\
&\quad + C\|Pu\|_{2\delta}^2 + C\|u\|_0^2 \\
&\leq \|Lu\|^2 + \sum_{j>0} |\langle X_j Pu \mid T^{2\delta} X_j u \rangle| \\
&\quad + C\|Pu\|_{2\delta}^2 + C\|u\|_0^2 + C\|Pu\|_{2\delta}\|X_j u\| \\
&\leq C \left(\|Lu\|^2 + \sum_{j>0} \|X_j Pu\|_{2\delta}^2 + \|Pu\|_{2\delta}^2 + \|u\|_0^2 \right) .
\end{aligned}
$$

It remains to treat $\|X_j Pu\|_{2\delta}^2$. We claim that

$$
\|X_j Pu\|_{2\delta}^2 \leq C \left(\|Lu\|_0^2 + \|Pu\|_{4\delta}^2 + \|u\|_0^2 \right) . \tag{2.16}
$$

We have indeed

$$
\begin{aligned}
\|X_j Pu\|_{2\delta}^2 &= \|\Lambda^{2\delta} X_j Pu\|^2 \\
&\leq C \left(\sum_{j>0} \|X_j \Lambda^{2\delta} Pu\|^2 + \|Pu\|_{2\delta}^2 \right) .
\end{aligned}
$$

Then using (2.8), we get[2]:

$$
\begin{aligned}
\sum_{j>0} \|X_j \Lambda^{2\delta} Pu\|^2 &\leq C \left(|\langle L\Lambda^{2\delta} Pu \mid \Lambda^{2\delta} Pu \rangle| + \|Pu\|_{2\delta}^2 + \|u\|_0^2 \right) \\
&\leq C \left(|\langle [L, \Lambda^{2\delta} P]u \mid \Lambda^{2\delta} Pu \rangle| + \|Pu\|_{4\delta}^2 + \|Lu\|_0^2 + \|u\|_0^2 \right) \\
&\leq C \left(\sum_j |\langle X_j u \mid Q^{4\delta} Pu \rangle| + \|Pu\|_{4\delta}^2 + \|Lu\|_0^2 + \|u\|_0^2 \right) \\
&\leq C \left(\|Pu\|_{4\delta}^2 + \|Lu\|_0^2 + \|u\|_0^2 \right) .
\end{aligned}
$$

This proves (2.16).
Taking $\delta \leq \min(\frac{\epsilon}{4}, \frac{1}{4})$, the right hand side is controlled.
The treatment of the term $|\langle PX_0 u \mid T^{2\delta} u \rangle|$ is similar.

Remark 2.14.
If $p = n$, the operator $\sum_j X_j^2 + \sum_j Y_j^2 + it \sum [X_j, Y_j]$ for $|t| < 1$ is also hypoelliptic. The problem is that this is the case $t = \pm 1$ which we would like to understand better.

---

[2] We cheat a little because we do not take care of the supports, but the pseudo-local character of the pseudo-differential operators permits to circumvent this problem. Here we recall that a linear operator P, which is defined on distributions, is pseudolocal if $\psi P \phi$ can be defined by a C^∞ distribution kernel, when ϕ and ψ are C^∞ functions with disjoint compact supports. A differential linear operator has evidently this property because $\psi P \phi$ is identically 0.

3

Compactness Criteria for the Resolvent of Schrödinger Operators

3.1 Introduction

It is well known [ReSi] that a Schrödinger operator, defined on $C_0^\infty(\mathbb{R}^d)$ by $-\Delta + V$, where V is semi-bounded from below on $\mathbb{R}^d$ and in $C^\infty(\mathbb{R}^d)$, admits a unique selfadjoint extension on $L^2(\mathbb{R}^d)$, i. e. is essentially self-adjoint. It is less known but still true that it is also the case under the weaker condition that $-\Delta + V$ is semi-bounded from below on C_0^∞ (see [Sim1] or for example [Hel11]), i.e. satisfying:

$$\exists C > 0, \ \forall u \in C_0^\infty(\mathbb{R}^d), \quad \langle(-\Delta + V)u \mid u\rangle \geq -C \|u\|^2 \ .$$

If in addition the potential $V(x)$ tends to $+\infty$ as $|x| \to \infty$, then the Schrödinger operator has a compact resolvent. The form domain of the operator is indeed given by $D_Q = \{u \in H^1(\mathbb{R}^d) \mid \sqrt{V + C_1}u \in L^2(\mathbb{R}^d)\}$ and it is immediate to verify, by a precompactness characterization, that the injection of D_Q into $L^2(\mathbb{R}^d)$ is compact. Our aim here is to analyze some cases when V does not necessarily tend to ∞.

The first well known example of such an operator which has nevertheless a compact resolvent is the operator $-\Delta + x_1^2 x_2^2$ in two dimension. One easy proof is as follow. Although the potential $V = x_1^2 x_2^2$ is 0 along $\{x_1 = 0\}$ or $\{x_2 = 0\}$, the estimate for the one-dimensional rescaled harmonic oscillator gives

$$-\Delta + x_1^2 x_2^2 \geq \frac{1}{2}\left(-\partial_{x_1}^2 + x_2^2 x_1^2\right) + \frac{1}{2}\left(-\partial_{x_2}^2 + x_1^2 x_2^2\right) \geq \frac{1}{2}(|x_2| + |x_1|) \ ,$$

where this comparison is the comparison between symmetric operators on $C_0^\infty(\mathbb{R}^2)$.

This permits to show that the form domain of the Schrödinger operator is included in the space $\{u \in H^1(\mathbb{R}^2) \mid |x|^{\frac{1}{2}}u \in L^2(\mathbb{R}^2)\}$, which is compactly embedded in $L^2(\mathbb{R}^2)$. Hence, the operator $-\Delta + x_1^2 x_2^2$ has a compact resolvent. This example can actually be treated by many approaches (see [Rob2], [Sim1], [HelNo3] and [HelMo]).

3.2 About Witten Laplacians and Schrödinger Operators

Let us consider the Laplacian introduced in (1.2)

$$\Delta_{\Phi}^{(0)} := -\Delta + |\nabla\Phi|^2 - \Delta\Phi .$$

For a C^∞ potential Φ on $\mathbb{R}^d$, this Laplacian is first defined as the Friedrichs extension associated with the form

$$\forall u \in C_0^\infty(\mathbb{R}^d), \langle u \mid \Delta_{\Phi}^{(0)} u \rangle = \left\| e^{-\Phi} \nabla_x e^{\Phi} u \right\|_{L^2(\mathbb{R}^d)}^2 .$$

Of course this is nothing but a specific Schrödinger operator and one can first think that it is enough to apply the general criteria for Schrödinger operators. Actually we look for criteria involving as directly as possible the function Φ.

This operator is called Witten Laplacian on 0-forms because it is a restriction of a more general Laplacian defined on all C^∞ forms, but it can also be considered as a Laplacian associated to a Dirichlet form like in probability. This Laplacian, which is positive by construction, is essentially self-adjoint on C_0^∞–which means admits a unique self-adjoint extension–(see for example[1] [Hel11] and [Sima]) and its self-adjoint closure has the domain

$$D(\Delta_{\Phi}^{(0)}) = \left\{ u \in L^2(\mathbb{R}^d),\ \Delta_{\Phi}^{(0)} u \in L^2(\mathbb{R}^d) \right\} .$$

Of course, it is easy to show that $\Delta_{\Phi}^{(0)}$ has a compact resolvent when

$$|\nabla\Phi(x)|^2 - \Delta\Phi(x) \to +\infty ,\ \text{as}\ |x| \to +\infty . \tag{3.1}$$

But this condition is not optimal ! A first improvement can indeed be obtained through the following " bracket argument".
We start from the inequality:

$$\sum_{j=1}^{d} \left(\|X_j u\|^2 + \alpha \|Y_j u\|^2 \right) \geq \pm i\sqrt{\alpha} \sum_{j=1}^{d} \langle [X_j, Y_j] u \mid u \rangle ,\ \forall u \in C_0^\infty(\mathbb{R}^d) , \tag{3.2}$$

where $X_j = \frac{1}{i} \partial_{x_j}$ and $Y_j = \partial_{x_j} \Phi$.
We observe also that

$$\sum_{j=1}^{d} [X_j, Y_j] = i\Delta\phi$$

and that

---

[1] We will present a similar argument in the analysis of the maximal accretivity of the Fokker-Planck operator (Section 5.2). Let us simply recall that the point is to show that $I + \Delta_{\Phi}^{(0)}$ has dense range in $L^2(\mathbb{R}^d)$.

$$\Delta_\Phi^{(0)} = \sum_{j=1}^d \left(X_j^2 + Y_j^2 - i[X_j, Y_j] \right) .$$

By convex combination, we obtain:

$$\langle \Delta_\Phi^{(0)} u \mid u \rangle \geq \int \left((1 - \epsilon)|\nabla\Phi|^2 + (\sqrt{\epsilon} - 1)\Delta\Phi \right) |u(x)|^2 \, dx , \quad \forall u \in C_0^\infty(\mathbb{R}^d) .$$
(3.3)

This gives:

$$\langle \Delta_\Phi^{(0)} u \mid u \rangle \geq (1 - \sqrt{\epsilon}) \int \left((1 + \sqrt{\epsilon})|\nabla\Phi|^2 - \Delta\Phi \right) |u(x)|^2 \, dx ,$$
(3.4)

for any $\epsilon \in]0, 1[$.

So we have obtained the following proposition (see [BoDaHel], [Hel11]).

Proposition 3.1.
Let us assume that there exists $t \in]0, 2[$ such that

$$t|\nabla\Phi(x)|^2 - \Delta\Phi(x) \to +\infty , \quad as \ |x| \to +\infty .$$
(3.5)

Then the Witten Laplacian $\Delta_\Phi^{(0)}$ has a compact resolvent.

One should notice that, for the function

$$\mathbb{R}^2 \ni (x_1, x_2) \mapsto \Phi(x_1, x_2) = x_1^2 x_2^2 + \varepsilon(x_1^2 + x_2^2) ,$$

where $\varepsilon \geq 0$, the corresponding potential $V = |\nabla\Phi|^2 - \Delta\Phi$ goes to $-\infty$ as $x_1 \to +\infty$ and $x_2 = 0$. Meanwhile the operator $\Delta_\Phi^{(0)}$ is positive by construction and we shall show in Theorem 11.10 that it has a compact resolvent if (and only if) $\varepsilon > 0$.

Remark 3.2.
One can also find criteria taking into account higher derivatives of Φ. See [BoDaHel] and Chapter 10.

Proposition 3.3.
If $\Delta_\Phi^{(0)}$ has a compact resolvent then the operator $S_\Phi := -\Delta + |\nabla\Phi|^2$ has also a compact resolvent.

Proof.
This follows immediately from the comparison:

$$0 \leq \Delta_\Phi^{(0)} \leq 2S_\Phi ,$$
(3.6)

between symmetric operators on $C_0^\infty(\mathbb{R}^d)$ and from the essential self-adjointness of these operators.

3.3 Compact Resolvent and Magnetic Bottles

Here we follow the proof of Helffer-Mohamed [HelMo], actually inspired by Kohn's proof presented in Section 2.3. We will analyze the problem for the family of operators:

$$P_A = \sum_{j=1}^{n}(D_{x_j} - A_j(x))^2 + \sum_{\ell=1}^{p} V_\ell(x)^2 \ . \tag{3.7}$$

Here the magnetic potential $A(x) = (A_1(x), A_2(x), \cdots, A_n(x))$ is supposed to be C^∞ and the electric potential $V(x) = \sum_j V_j(x)^2$ is such that $V_j \in C^\infty$. Under these conditions, the operator is essentially self-adjoint on $C_0^\infty(\mathbb{R}^n)$. We note also that it has the form:

$$P_A = \sum_{j=1}^{n+p} X_j^2 = \sum_{j=1}^{n} X_j^2 + \sum_{\ell=1}^{p} Y_\ell^2 \ ,$$

with

$$X_j = (D_{x_j} - A_j(x)), \ j = 1, \ldots, n \ , \ Y_\ell = V_\ell \ , \ \ell = 1, \ldots, p \ .$$

In particular, the magnetic field is recovered by observing that

$$B_{jk} = \frac{1}{i}[X_j, X_k] = \partial_j A_k - \partial_k A_j \ , \ \text{ for } j, k = 1, \ldots, n \ .$$

We start with two trivial easy cases.
First we consider the case when $V \to +\infty$. In this case, it is well known that the operator has a compact resolvent.(see the argument below).
On the opposite, we assume that $V = 0$ and consider the case when $n = 2$ and when $V = 0$. We assume moreover that $B(x) = B_{12} \geq 0$. Then one immediately observe the following inequality:

$$\int B(x)|u(x)|^2 dx \leq ||X_1 u||^2 + ||X_2 u||^2 = \langle P_A u \mid u \rangle \ . \tag{3.8}$$

Under the condition that $\lim_{x \to \infty} B(x) = +\infty$, this implies that the operator has a compact resolvent .

Example 3.4.

$$A_1(x_1, x_2) = x_2 x_1^2 \ , \ A_2(x_1, x_2) = -x_1 x_2^2 \ .$$

Indeed it is sufficient to show that the form domain of the operator $D(q_A)$ which is defined by:

$$D(q_A) = \{u \in L^2(\mathbb{R}^n) \ , \ X_j u \in L^2(\mathbb{R}^n) \ , \ \text{ for } j = 1, \ldots, n + p\} \ . \tag{3.9}$$

is contained in the weighted L^2-space,

$$L^2_\rho(\mathbb{R}^n) = \{u \in \mathcal{S}'(\mathbb{R}^n) \mid \rho^{\frac{1}{2}}u \in L^2(\mathbb{R}^n)\} , \tag{3.10}$$

for some positive continuous function $x \mapsto \rho(x)$ tending to ∞ as $|x| \to \infty$.
In order to treat more general situations, we introduce the quantities:

$$m_q(x) = \sum_\ell \sum_{|\alpha|=q} |\partial_x^\alpha V_\ell| + \sum_{j<k} \sum_{|\alpha|=q-1} |\partial_x^\alpha B_{jk}(x)| . \tag{3.11}$$

It is easy to reinterpret this quantity in terms of commutators of the X_j's.
When $q = 0$, the convention is that

$$m_0(x) = \sum_\ell |V_\ell(x)| . \tag{3.12}$$

Let us also introduce

$$m^r(x) = 1 + \sum_{q=0}^r m_q(x) . \tag{3.13}$$

Then the criterion is

Theorem 3.5.
Let us assume that there exists r and a constant C such that

$$m_{r+1}(x) \leq C\, m^r(x) , \quad \forall x \in \mathbb{R}^n , \tag{3.14}$$

and

$$m^r(x) \to +\infty , \quad as \; |x| \to +\infty . \tag{3.15}$$

Then $P_A(h)$ has a compact resolvent.

Remark 3.6.
It is shown in [Mef], that one can get the same result as in Theorem 3.5 under
the weaker assumption that

$$m_{r+1}(x) \leq Cm^r(x)^{1+\delta} , \tag{3.16}$$

where $\delta = \frac{1}{2^{r+1}-3}$ ($r \geq 1$). This result is optimal for $r = 1$ according to a coun-
terexample by A. Iwatsuka [Iw]. He gives indeed an example of a Schrödinger
operator which has a non compact resolvent and such that $\sum_{j<k} |\nabla B_{jk}(x)|$
has the same order as $\sum_{j<k} |B_{j<k}|^2$.
Other generalizations are given in [She] (Corollary 0.11) (see also references
therein and [KonShu] for a quite recent contribution including other references).
One can for example replace $\sum_j V_j^2$ by V and the conditions on the m_j's can
be reformulated in terms of the variation of V and B in suitable balls. In
particular A. Iwatsuka [Iw] showed that a necessary condition is:

$$\int_{B(x,1)} \left(V(x) + \sum_{j<k} B_{jk}(x)^2 \right) dx \to +\infty \; \text{as} \; |x| \to +\infty , \tag{3.17}$$

where $B(x,1)$ is the ball of radius 1 centered at x.

Remark 3.7.
If $p = n$, the operator $\sum_{j=1}^n X_j^2 + \sum_{j=1}^n Y_j^2 + it \sum [X_j, Y_j]$, for $|t| < 1$, has also a compact resolvent under the conditions of Theorem 3.5. The problem is that this is the case $t = \pm 1$ which appears in the analysis of the Witten Laplacian.

Before entering into the core of the proof, we observe that we can replace $m^r(x)$ by an equivalent C^∞ function $\Psi(x)$ which has the property that there exist constants C_α and $C > 0$ such that:

$$\begin{aligned} \tfrac{1}{C}\Psi(x) \le m^r(x) \le C\Psi(x) \,, \\ |D_x^\alpha \Psi(x)| \le C_\alpha \Psi(x) \,. \end{aligned} \tag{3.18}$$

Indeed, it suffices to replace quantities like $\sum |u_k|$ by $(\sum |u_k|^2)^{1/2}$, in the definition (3.11) of m_q. Tne second condition is a consequence of (3.14). In the same spirit as in Kohn's proof, let us introduce for all $s > 0$

Definition 3.8.
We denote by M^s the space of C^∞ functions T such that there exists C_s such that:

$$||\Psi^{-1+s}Tu||^2 \le C_s \left(\langle P_A u \mid u \rangle + ||u||^2 \right) \,, \quad \forall u \in C_0^\infty(\mathbb{R}^n) \,. \tag{3.19}$$

We observe that

$$V_\ell \in M^1 \,, \tag{3.20}$$

and we will show the

Lemma 3.9.
$$[X_j, X_k] \in M^{\frac{1}{2}} \,, \quad \forall j, k = 1, \ldots, n \,. \tag{3.21}$$

Another claim is contained in the

Lemma 3.10.
If T is in M^s and $|\partial_x^\alpha T| \le C_\alpha \Psi$ then $[X_k, T] \in M^{\frac{s}{2}}$, when $|\alpha| = 1$ or $|\alpha| = 2$.

Assuming these two lemmas, then it is clear that

$$\Psi(x) \in M^{2^{-r}} \,.$$

Lemma 3.10 and (3.20) lead to

$$\partial_x^\alpha V_\ell \in M^{2^{-|\alpha|}} \,,$$

and we deduce from Lemmas 3.9 and 3.10:

$$\partial_x^\alpha B_{jk} \in M^{2^{-(|\alpha|+1)}} \,.$$

The proof of Theorem 3.5 then becomes easy.

Proof of Lemma 3.9

We start from the identity (and observing that $X_j^* = X_j$):

$$
\begin{aligned}
||\Psi^{-\frac{1}{2}}[X_j, X_k]u||^2 &= \langle (X_j X_k - X_k X_j)u \mid \Psi^{-1}[X_j, X_k]u \rangle \\
&= \langle X_k u \mid X_j \Psi^{-1}[X_j, X_k]u \rangle \\
&\quad - \langle X_j u \mid X_k \Psi^{-1}[X_j, X_k]u \rangle \\
&= \langle X_j u \mid \Psi^{-1}[X_k, X_j]X_k u \rangle \\
&\quad - \langle X_k u \mid \Psi^{-1}[X_k, X_j]X_k u \rangle \\
&\quad + \langle X_j u \mid [X_k, \Psi^{-1}[X_k, X_j]]u \rangle \\
&\quad - \langle X_k u \mid [X_j, \Psi^{-1}[X_k, X_j]]u \rangle .
\end{aligned}
$$

If we observe that $\Psi^{-1}[X_k, X_j]$ and $[X_k, \Psi^{-1}[X_k, X_j]]$ are bounded (look at the definition of Ψ), we obtain:

$$
||\Psi^{-\frac{1}{2}}[X_j, X_k]u||^2 \leq C \left(||X_k u||^2 + ||X_j u||^2 + ||u||^2 \right) .
$$

This ends the proof of the lemma.

Proof of Lemma 3.10

Let $T \in M^s$. For each k, we can write:

$$
\begin{aligned}
||\Psi^{-1+\frac{s}{2}}[X_k, T]u||^2 &= \langle \Psi^{-1+s}(X_k T - T X_k)u \mid \Psi^{-1}[X_k, T]u \rangle \\
&= \langle \Psi^{-1+s}X_k T u \mid \Psi^{-1}[X_k, T]u \rangle \\
&\quad - \langle \Psi^{-1+s}T X_k u \mid \Psi^{-1}[X_k, T]u \rangle \\
&= \langle \Psi^{-1+s}T u \mid \Psi^{-1}[X_k, T]X_k u \rangle \\
&\quad - \langle X_k u \mid \Psi^{-1}[X_k, T]\Psi^{-1+s}T u \rangle \\
&\quad + \langle T u \mid [X_k, \Psi^{-2+s}[X_k, T]]u \rangle \\
&= \langle \Psi^{-1+s}T u \mid \Psi^{-1}[X_k, T]X_k u \rangle \\
&\quad - \langle X_k u \mid \Psi^{-1}[X_k, T]\Psi^{-1+s}T u \rangle \\
&\quad + \langle \Psi^{-1+s}T u \mid \Psi^{1-s}[X_k, \Psi^{-2+s}[X_k, T]]u \rangle .
\end{aligned}
$$

We now observe, according to the assumptions of the lemma and the properties of Ψ, that $\Psi^{1-s}[X_k, \Psi^{-2+s}[X_k, T]]$ and $\Psi^{-1}[X_k, T]$ are bounded.

So finally we get:

$$
||\Psi^{-\frac{1}{2}}[X_j, T]u||^2 \leq C \left(||\Psi^{-1+s}T u||^2 + ||X_k u||^2 + ||u||^2 \right) .
$$

This ends the proof of the lemma.

Remark 3.11.

Helffer-Mohamed describe also in [HelMo] the essential spectrum when the compactness criterion of the resolvent is not satisfied.

We mention also the negative answer to the problem of finding magnetic bottles for the Dirac operator due to Helffer-Nourrigat-Wang [HeNoWa] (see the book by B. Thaller [Tha] on this question). It is indeed "essentially" (the proof is under additional technical conditions) shown that, in the two dimensional case, the resolvent of the Dirac operator $\sum_{j=1}^{2} \sigma_j(D_{x_j} - A_j(x))$ is never compact. Here the σ_j are two by two self-adjoint matrices such that

$$\sigma_1^2 = \sigma_2^2 = I \ , \ \sigma_1 \sigma_2 = -\sigma_2 \sigma_1 \ .$$

The standard choice is

$$\sigma_1 = \begin{pmatrix} 0 & 1 \\ 1 & 0 \end{pmatrix} \ , \ \sigma_2 = \begin{pmatrix} 0 & -i \\ i & 0 \end{pmatrix} \ .$$

We also observe that the square of this operator is diagonal and that the diagonal corresponds to the so called Pauli operators

$$\sum_{j=1}^{2} (D_{x_j} - A_j(x))^2 \pm B(x) \ .$$

These operators have the structure observed in Remark 3.7, with $t = \pm 1$.

Remark 3.12.
As it can for example be seen in [HelNo3], similar problems occur in the theory of the $\bar{\partial}$-Neumann Laplacian and more specifically for the $\Box_b$ operator. We refer to the quite recent papers by Fu-Straube [FuSt] and Christ-Fu [ChFu] for a presentation of the theory initiated by J. Kohn [Ko] and for a complete list of references.

4

Global Pseudo-differential Calculus

This chapter is a review, in a specific case, of basic properties of pseudo-differential operators. Our motivation was the construction of a chain of powers of positive "elliptic" operators which could replace the chain $(1 - \Delta)^s$ ($s \in \mathbb{R}$) appearing in Kohn's proof of hypoellipticity. Because this leads, independently of –but motivated by– this application, to interesting questions about the Weyl-Hörmander calculus, we have added new results on this calculus, with the kind help of J.M. Bony. In a first reading, the reader which is not a specialist in microlocal analysis can skip part of these techniques and proceeds by admitting the result of Theorem 4.8. Note that similar results were obtained under stronger assumptions in the seventies (see the comments at the end of the chapter). The main properties of these pseudo-differential operators will be recalled in Section 4.2.

4.1 The Weyl-Hörmander Pseudo-differential Calculus

We just give in this section, a small account on the so-called Weyl-Hörmander calculus. It is in some sense the most sophisticated and the most powerful version of the pseudo-differential calculus[1], whose first version was presented around (2.12) in Section 2.3.

In $\mathbb{R}^{2d}_{z,\zeta}$ (this will be applied later with $d = 2n$, $z = (x,v)$ and $\zeta = (\xi,\eta)$) we consider the class of $\mathcal{C}^\infty$ functions which satisfy

$$\forall \alpha, \beta \in \mathbb{N}^d, \quad \exists C_{\alpha,\beta} > 0, \forall (z,\zeta) \in \mathbb{R}^{2d}, \quad \left| \partial_z^\alpha \partial_\zeta^\beta a(z,\zeta) \right| \leq C_{\alpha,\beta} \Psi(z,\zeta)^{m-|\beta|},$$

for some $m \in \mathbb{R}$.

The function Ψ is a fixed $\mathcal{C}^\infty$ function bounded from below by 1, with other properties specified below. By introducing the metric

---

[1] resulting of the efforts of many mathematicians mainly in the period 70-85, including R. Beals and L. Hörmander,

$$g = dz^2 + \frac{d\zeta^2}{\Psi^2},$$

the above condition writes $|T_1 \cdots T_J a(z,\zeta)| \leq C_J \Psi^m(z,\zeta)$ for any finite sequence of vector fields $(T_j)_{j=1,\ldots,J}$ such that $g(T_j) \leq 1$ uniformly. This class of symbols is usually denoted by $S(\Psi^m, g)$ and we shall write shortly S_Ψ^m. This space of symbols S_Ψ^m endowed with the seminorms

$$|a|_{k,S_\Psi^m} = \sup_{|\alpha+\beta| \leq k} \sup_{(z,\zeta) \in \mathbb{R}^{2d}} \Psi^{-m+|\beta|}(z,\zeta) \left| \partial_z^\alpha \partial_\zeta^\beta a(z,\zeta) \right|, \quad k \in \mathbb{N}$$

is a Fréchet space.

For any symbol in $\mathcal{S}'(\mathbb{R}^{2d})$, the pseudo-differential operator $a^W(z, D_z)$ is an operator from $\mathcal{S}(\mathbb{R}^d)$ into $\mathcal{S}'(\mathbb{R}^d)$ whose Schwartz kernel (that is distribution kernel) is defined by the oscillatory integral:

$$a^W(z, D_z)(z, z') = \frac{1}{(2\pi)^d} \int_{\mathbb{R}^d} e^{i(z-z') \cdot \zeta} a\left(\frac{z+z'}{2}, \zeta\right) d\zeta. \qquad (4.1)$$

This is known as the Weyl quantization of the symbol a and other quantizations, as we have seen in Section 2.1 (see (2.12)) are possible

$$a^{(t)}(z, D_z)(z, z') = \frac{1}{(2\pi)^d} \int_{\mathbb{R}^d} e^{i(z-z') \cdot \zeta} a((1-t)z + tz', \zeta) d\zeta, \quad t \in [0, 1], \qquad (4.2)$$

where $t = 0$ corresponds to the standard pseudo-differential calculus and $t = 1$ to the adjoint calculus. The Weyl quantization corresponds to the case $t = 1/2$ and has the following nice property.

Proposition 4.1.
The operator $a^W(z, D_z)$ is symmetric[2] on $\mathcal{S}(\mathbb{R}^d)$, when a is real.

Its central role in the theory, is due to the fact that it exhibits the fundamental relationship between quantization and the symplectic structure of $\mathbb{R}^{2d} = T^*\mathbb{R}^d$ endowed with its canonical symplectic form

$$\sigma = \sum_{j=1}^d d\zeta_j \wedge dz_j, \quad \sigma(Z, Z') = \sum_{j=1}^d (\zeta_j z_j' - z_j \zeta_j'). \qquad (4.3)$$

Here and in the sequel, the capital character Z denotes the pair (z, ζ) in $\mathbb{R}^{2d}$.
  The dual metric with respect to the symplectic form σ is given here by

---

[2] This means:

$$\langle a^W(z, D_z)u \mid v \rangle_{L^2(\mathbb{R}^d)} = \langle u \mid a^W(z, D_z)v \rangle_{L^2(\mathbb{R}^d)}, \quad \forall u, v \in \mathcal{S}(\mathbb{R}^d).$$

$$g^\sigma = \Psi^2 dz^2 + d\zeta^2 .$$

The condition $\Psi \geq 1$ ensures that the metric is compatible with the uncertainty principle, which takes here the form:

$$g \leq g^\sigma . \tag{4.4}$$

We now assume that the function Ψ satisfies uniformly for some constants $c_1 \geq 1$, $c_2 \geq 0$ and $\nu \geq 0$:

$$\left. \begin{array}{l} |z - z'| \leq c_1^{-1} \\ |\zeta - \zeta'| \leq c_1^{-1}\Psi(z,\zeta) \end{array} \right\} \Rightarrow \left(\frac{\Psi(z,\zeta)}{\Psi(z',\zeta')} \right)^{\pm 1} \leq c_1 \tag{4.5}$$

and

$$\left(\frac{\Psi(z,\zeta)}{\Psi(z',\zeta')} \right)^{\pm 1} \leq c_2 \left(1 + \Psi(z,\zeta)^2 |z - z'|^2 + |\zeta - \zeta'|^2 \right)^\nu . \tag{4.6}$$

Then the metric g satisfies the Hörmander **slowness condition**:

$$(g_Z(Z - Z') \leq C_1) \Rightarrow \left(\sup_{T \neq 0} \left(\frac{g_Z(T)}{g_{Z'}(T)} \right)^{\pm 1} \leq \frac{1}{C_1} \right) \tag{4.7}$$

for some uniform constant $C_1 \leq 1$, and the Hörmander **temperance condition**:

$$\left(\sup_{T \neq 0} \frac{g_Z(T)}{g_{Z'}(T)} \right)^{\pm 1} \leq C_2 \left(1 + g_Z^\sigma(Z - Z') \right)^N , \tag{4.8}$$

for some uniform constants $C_2 > 0$ and $N > 0$.
It is possible to check that the temperance is equivalent to the symmetric temperance introduced in [BoLe] and used in [BonChe] and [NaNi].

4.2 Basic Properties

We now give the consequences of these properties, which can be found in [Hor2]-Chap XVIII after noticing the value of the gain for our specific calculus:

$$\lambda(Z) = \left(\min_{T \neq 0} \frac{g_Z^\sigma(T)}{g_Z(T)} \right)^{1/2} = \Psi(Z) .$$

4.2.1 Composition

For any $m \in \mathbb{R}$ and any $a \in S_\Psi^m$ the operator $a^W(x, D_x)$ acts continuously on $\mathcal{S}(\mathbb{R}^d)$ and on $\mathcal{S}'(\mathbb{R}^d)$. Thus two pseudo-differential operators can be composed as operators in $\mathcal{L}(\mathcal{S}(\mathbb{R}^d))$ or in $\mathcal{L}(\mathcal{S}'(\mathbb{R}^d))$. The natural question is then whether the product is also a pseudo-differential operator. This is the topic of the next subsection.

4.2.2 The Algebra $\cup_{m\in\mathbb{R}} \operatorname{Op} S_{\Psi}^{m}$

If $a \in S^{m}(\mathbb{R}^{2d})$ and $b \in S^{m'}(\mathbb{R}^{d})$, then $a^{W}(z, D_z) \circ b^{W}(z, D_z) = c^{W}(z, D_z)$ with $c \in S_{\Psi}^{m+m'}$. The symbol c is denoted by $a\sharp^{W} b$ and we have the expansion:

$$a\sharp^{W} b(Z) = \left(e^{\frac{i}{2}\sigma(D_{Z_1}, D_{Z_2})} a(Z_1) b(Z_2)\right)\Big|_{Z_1=Z_2=Z} \tag{4.9}$$

$$= \sum_{j=0}^{J-1} \frac{\left(\frac{i}{2}\sigma(D_{Z_1}, D_{Z_2})\right)^{j}}{j!} a(Z_1) b(Z_2)\Big|_{Z_1=Z_2=Z} \tag{4.10}$$

$$+ \int_{0}^{1} \frac{(1-\theta)^{J-1}}{(J-1)!} e^{\frac{i}{2}\theta\sigma(D_{Z_1}, D_{Z_2})} \left(\frac{i}{2}\sigma(D_{Z_1}, D_{Z_2})\right)^{J} a(Z_1) b(Z_2)\Big|_{Z_1=Z_2=Z} \tag{4.11}$$

$$= \sum_{j=0}^{J-1} \frac{\left(\frac{i}{2}\sigma(D_{Z_1}, D_{Z_2})\right)^{j}}{j!} a(Z_1) b(Z_2)\Big|_{Z_1=Z_2=Z} + R_J(a,b)(Z), \tag{4.12}$$

where R_J is a continuous bilinear operator from $S_{\Psi}^{m} \times S_{\Psi}^{m'}$ into $S_{\Psi}^{m+m'-J}$ (i.e. any seminorm of $R_J(a,b)$ is controlled by some bilinear expression of a finite number of seminorms of a and b).

Let $\operatorname{Op} S_{\Psi}^{m}$ denote the set of pseudo-differential operators, the main consequences of the previous relations can be summarized by:

Proposition 4.2.
The space $\cup_{m\in\mathbb{R}} \operatorname{Op} S_{\Psi}^{m}$ is an algebra with

$$\operatorname{Op} S_{\Psi}^{m} \circ \operatorname{Op} S_{\Psi}^{m'} \subset \operatorname{Op} S_{\Psi}^{m+m'} \tag{4.13}$$

and

$$\left[\operatorname{Op} S_{\Psi}^{m}, \operatorname{Op} S_{\Psi}^{m'}\right] \subset \operatorname{Op} S_{\Psi}^{m+m'-1}. \tag{4.14}$$

Note also that the principal symbols (symbol modulo lower order terms[3]) of $a^{W}(z, D_z) \circ b^{W}(z, D_z)$ and $i[a^{W}(z, D_z), b^{W}(z, D_z)]$ respectively equal ab and the Poisson bracket $\{a, b\} = \sum_{k=1}^{d} \partial_{\zeta_k} a \partial_{z_k} b - \partial_{z_k} a \partial_{\zeta_k} b$.

4.2.3 Equivalence of Quantizations

Since our metric g is splitted, $g_Z(t_z, -t_\zeta) = g_Z(t_z, t_\zeta)$, all the quantizations are equivalent at the principal symbol level. More precisely, for any $a \in S_{\Psi}^{m}$ and for any $t, t' \in [0, 1]$, there exists a unique symbol $a_{t,t'} \in S_{\Psi}^{m}$ such that $a_{t,t'}^{(t')}(z, D_z) = a^{(t)}(z, D_z)$. They satisfy

$$a_{t,t'} = e^{-i(t-t')D_z \cdot D_\zeta} a = \sum_{j=0}^{J-1} \frac{(-i(t-t')D_z D_\zeta)^{j}}{j!} a + R_{t,t',J}(a),$$

where $R_{t,t',J}$ is a continuous operator from S_{Ψ}^{m} into S_{Ψ}^{m-J}. Hence in any quantization the symbol of the formal adjoint of $a^{t}(z, D_Z)$ is $\bar{a}$ up to lower order terms.

---

[3] Here the case $\Phi = 1$ is not excluded and "lower order" may mean "same order".

4.2.4 $L^2(\mathbb{R}^d)$-Continuity

The following theorem is an extension of the celebrated Calderon-Vaillancourt Theorem giving the L^2-continuity of pseudo-differential operator of order 0:

Theorem 4.3.

$$\mathrm{Op}\, S_\Psi^0 \subset \mathcal{L}(L^2(\mathbb{R}^d)). \tag{4.15}$$

According to the previous remark, this holds for any quantization.

4.2.5 Compact Pseudo-differential Operators

Proposition 4.4.
If the function Ψ satisfies $\lim_{(z,\zeta)\to\infty} \Psi(z,\zeta) = +\infty$ then, for any $\varepsilon > 0$, $\mathrm{Op}\, S_\Psi^{-\varepsilon}$ is continuously embedded in the space $\mathcal{K}(L^2(\mathbb{R}^d))$ of compact operators in $L^2(\mathbb{R}^d)$.

4.3 Fully Elliptic Operators and Beals Type Characterization

A pseudo-differential operator $a^W(z, D_z) \in \mathrm{Op}\, S_\Psi^m$ is said to be fully elliptic if its symbol satisfies

$$|a(z,\zeta)| \geq C^{-1}\Psi^m(z,\zeta), \tag{4.16}$$

for some $C > 0$, while it is said elliptic if the inequality holds up to some remainder $R \in S_\Psi^{m-\delta}$, $\delta > 0$. Any fully elliptic operator admits at any order a left and right parametrix. We can first write

$$a \sharp^W a^{-1} = 1 - r_1, \quad \text{with } r_1 \in S_\Psi^{-1},$$

and then $\sharp^W$-multiply on the right with $1 + r_1$ in order to get $1 - r_2$, with $r_2 \in S_\Psi^{-2}$, at the right-hand side, and so on. The left parametrix at arbitrary order is obtained similarly.

In [BonChe], J.M. Bony and J.Y. Chemin introduced in a wide generality Sobolev spaces attached to the Weyl-Hörmander calculus and gave a version of the Beals criterion (some other details and improvements where given by J.M. Bony [Bon1] in his graduate course at Ecole Polytechnique in 1997-1998). In our case, the results of [BonChe] provide a Sobolev scale of Hilbert spaces H_Ψ^s, indexed by $s \in \mathbb{R}$, such that

$$\mathcal{S}(\mathbb{R}^d) \subset H_\Psi^s \subset H_\Psi^{s'} \subset \mathcal{S}'(\mathbb{R}^d), \text{ for } s \geq s',$$
$$\forall s, m \in \mathbb{R}, \forall a \in S_\Psi^m, a^W(z, D_z) \in \mathcal{L}\left(H_\Psi^s; H_\Psi^{s-m}\right),$$
$$H_\Psi^0 = L^2(\mathbb{R}^d) \quad \text{and} \quad (H_\Psi^s)' = H_\Psi^{-s}.$$

In our case, the metric is diagonal in a fixed basis (see [NaNi] for a detailed version of Remark 5.6 of [BonChe]) and an operator $A : \mathcal{S}(\mathbb{R}^d) \to \mathcal{S}'(\mathbb{R}^d)$ belongs to $\mathrm{Op}\, S_\Psi^m$ if and only if it satisfies the estimates

$$\forall \alpha, \beta \in \mathbb{N}^d, \exists C_{\alpha,\beta} > 0, \quad \left\| \mathrm{ad}\,_z^\alpha \mathrm{ad}\,_{D_z}^\beta A \right\|_{\mathcal{L}(H_\Psi^s ; H_\Psi^{s-m+|\alpha|})} \leq C_{\alpha,\beta} ,$$

for some $s \in \mathbb{R}$.

Moreover the maps $a \mapsto \left\| \mathrm{ad}\,_z^\alpha \mathrm{ad}\,_{D_z}^\beta A \right\|_{\mathcal{L}(H_\Psi^s ; H_\Psi^{s-m+|\alpha|})}$ (with $A = a^W(z, D_z)$) define, for $(\alpha, \beta) \in \mathbb{N}^d \times \mathbb{N}^d$, a set of seminorms which is equivalent to the set of semi-norms $a \mapsto \left(|a|_{k,S_\Psi^m} \right)_{k \in \mathbb{N}}$ on S_Ψ^m. We now apply these results in some specific case. Our aim is to check that arbitrary real powers of a positive elliptic pseudo-differential operator are pseudo-differential operators. Although some statements sound like standard results, these results have to be checked for general Ψ (see the comments in Section 4.5 below).

Proposition 4.5.
a) Let $A = a^W(z, D_z)$ be a fully elliptic operator in $\mathrm{Op}\, S_\Psi^m$,

$$|a| \geq C^{-1} \Psi^m .$$

If the operator $A : H_\Psi^{s_0} \to H_\Psi^{s_0-m}$ is invertible for some $s_0 \in \mathbb{R}$, then it is an isomorphism from H_Ψ^s onto H_Ψ^{s-m} for any $s \in \mathbb{R}$.
b) Let $A = a^W(z, D_z) \in \mathrm{Op}\, S_\Psi^m$, $m > 0$, be a symmetric (the symbol a is real valued) elliptic operator,

$$|a| \geq C^{-1} \Psi^m + R ,$$

with $R \in S_\Psi^{m-\delta}$, $\delta > 0$. Then the operator $(A, D(A) = H_\Psi^m)$ is self-adjoint on $L^2(\mathbb{R}^d)$ and the Sobolev scale $(H_\Psi^s)_{s \in \mathbb{R}}$ coincides with the Sobolev scale associated with the self-adjoint operator $(A, D(A) = H_\Psi^m)$.
Moreover, if $A \geq c_0 \mathrm{Id}$, then for any $s \in \mathbb{R}$ the norms $u \mapsto \|u\|_{H_\Psi^s}$ and $u \mapsto \left\| A^{s/m} u \right\|_{L^2}$ are equivalent.

Proof of a).
First note that A^* with symbol $\bar{a}$ satisfies the same properties as A with s_0 replaced by $-s_0 + m$. The result is proved if $A_s := A$ considered as an operator from H_Ψ^s into H_Ψ^{s-m} has a closed range, $\mathrm{Ker}\,(A_s) = \{0\}$ and $\mathrm{Ker}\,(A_s^*) = 0$. This is a consequence of elliptic regularity. For any $s \in \mathbb{R}$, $A \in \mathrm{Op}\, S_\Psi^m$ implies

$$\forall u \in H_\Psi^s, \|Au\|_{H_\Psi^{s-m}} \leq C \|u\|_{H_\Psi^s} .$$

Conversely, since A admits a left parametrix B_J at any order, i.e.

$$\exists B_J \in \mathrm{Op}\, S_\Psi^{-m}, B_J A = \mathrm{Id} + R_J , \quad \text{with } R_J \in \mathrm{Op}\, S_\Psi^{-J} ,$$

the estimate

$$\|v\|_{H^s} \leq \|B_J Au - R_J u\|_{H^s} \leq C \|Av\|_{H_{\Psi}^{s-m}} + \|u\|_{H_{\Psi}^{s-m}}$$

holds for any $v \in H_{\Psi}^s$ (take $J \geq m$). Hence the operator $A_s : H_{\Psi}^s \to H_{\Psi}^{s-m}$ has a closed range.

If $u \in H_{\Psi}^s$ satisfies $A_s u = 0$, there are two cases:

1. $s \geq s_0$.
   Then $u \in H_{\Psi}^{s_0}$ belongs to $\mathrm{Ker}\,(A)$ and we have $u = 0$.
2. $s < s_0$.
   Using again the left parametrix with J large enough, we get

$$0 = B_J Au = u + R_J u \,,$$

which implies that $u = -R_J u$ belongs to $H_{\Psi}^{s_0}$. Therefore $u = 0$.

Similarly the same properties for A^* lead to $\mathrm{Ker}\,(A_s^*) = \{0\}$ and A defines an isomorphism from H_{Ψ}^s onto H_{Ψ}^{s-m} for any $s \in \mathbb{R}$.

Proof of b).
Notice first that an elliptic real valued symbol a can be made fully elliptic by adding it_0 with $t_0 \in \mathbb{R}$ and $|t_0|$ large enough. Hence, it suffices to check that for $t_0 \in \mathbb{R}$, $|t_0|$ large enough, $it_0 + A : H_{\Psi}^m \to L^2(\mathbb{R}^d)$ is invertible and then to apply the results of a). The identity (4.10) shows that the remainder $R_J(a,b)$ depends only on the Jth derivatives of the symbols a and b and leads to

$$(it_0 + a)\sharp^W \left(\frac{1}{it_0 + a}\right) = 1 + \frac{1}{2i}\left\{it_0 + a, \frac{1}{it_0 + a}\right\} + R_2(it_0 + a, (it_0 + a)^{-1})$$

$$= 1 + R_2(a, (it_0 + a)^{-1}) \,, \tag{4.17}$$

where R_2 is continuous from $S_{\Psi}^m \times S_{\Psi}^{-m+2}$ to S_{Ψ}^0. The seminorms of the symbol $(it_0 + a)^{-1}$ in S_{Ψ}^{-m+2} are of order $|t_0|^{-2/m}$ and the right-hand side $1 + R_2(a, (t_0 + a)^{-1})$ is invertible in $\mathcal{L}(L^2(\mathbb{R}^d))$ for $|t_0|$ large enough. Hence $(it_0 + A) : H_{\Psi}^m \to L^2(\mathbb{R}^d)$ admits a right inverse. A left inverse is constructed similarly for $|t_0|$ large enough and $(A, D(A) = H_{\Psi}^m)$ is self-adjoint. The equality of the two Sobolev scales and the equivalence of the norms are consequences of a). $\blacksquare$

Remark 4.6.
An example of such an operator A is $t_0 + a(z, D_z)$ with $a \geq 0$, $a \in S_{\Psi}^1$ elliptic and the constant $t_0 > 0$ large enough.

Proposition 4.7.
Let $A \in \mathrm{Op}\, S_{\Psi}^m$, $m > 0$, be elliptic and satisfy $A \geq c_0 \mathrm{Id}$. Then for any fixed $\lambda \notin \sigma(A)$, $(A-\lambda)^{-1}$ belongs to $\mathrm{Op}\, S_{\Psi}^{-m}$. Moreover, the seminorms of $(t+A)^{-1}$ in $\mathrm{Op}\, S_{\Psi}^{-m}$ are bounded uniformly with respect to $t \geq 0$.

By Proposition 4.5, $(A, D(A) = H_\Psi^m)$ is self-adjoint and for any fixed $s \in \mathbb{R}$ the norms $\|u\|_{H_\Psi^s}$ and $\|A^{s/m}u\|_{L^2}$ are equivalent. We shall use the Beals criterion in the form

$$\forall \alpha, \beta \in \mathbb{N}^d, \exists C_{\alpha,\beta} > 0, \quad \left\| A^{1+|\alpha|/m} \text{ad}_z^\alpha \text{ad}_{D_z}^\beta (\lambda - A)^{-1} \right\|_{\mathcal{L}(L^2)} \le C_{\alpha,\beta,\lambda} .$$

The relation

$$0 = \text{ad}_z^\alpha \text{ad}_{D_z}^\beta ((A - \lambda)(A - \lambda)^{-1})$$

$$= \sum_{(\alpha_1,\beta_1) \le (\alpha,\beta)} c_{\alpha,\beta,\alpha_1\beta_1} \left[\text{ad}_z^{\alpha-\alpha_1} \text{ad}_{D_z}^{\beta-\beta_1}(A - \lambda) \right] \circ \left[\text{ad}_z^{\alpha_1} \text{ad}_{D_z}^{\beta_1}(A - \lambda)^{-1} \right]$$

makes sense, if one considers the factors $\text{ad}_z^{\alpha_1} \text{ad}_{D_z}^{\beta_1}(A - \lambda)^{-1}$ as continuous operators from $\mathcal{S}(\mathbb{R}^d)$ to $\mathcal{S}'(\mathbb{R}^d)$ because the pseudo-differential factors $\text{ad}_z^{\alpha-\alpha_1} \text{ad}_{D_z}^{\beta-\beta_1}(A - \lambda)$ are continuous in $\mathcal{S}'(\mathbb{R}^d)$.

The result is derived by induction from this relation in the form

$$\text{ad}_z^\alpha \text{ad}_{D_z}^\beta (A - \lambda)^{-1}$$

$$= - \sum_{(\alpha_1,\beta_1) < (\alpha,\beta)} c_{\alpha,\beta,\alpha_1\beta_1} \left[\text{ad}_z^{\alpha-\alpha_1} \text{ad}_{D_z}^{\beta-\beta_1} A \right] \circ \left[\text{ad}_z^{\alpha_1} \text{ad}_{D_z}^{\beta_1}(A - \lambda)^{-1} \right] ,$$

with:

$$A(A - \lambda)^{-1} \in \mathcal{L}(L^2) ,$$

$$A^{|\alpha|/m} \left[\text{ad}_z^{\alpha-\alpha_1} \text{ad}_{D_z}^{\beta-\beta_1} A \right] A^{-1-|\alpha_1|/m} \in \mathcal{L}(L^2) ,$$

$$\text{and } A^{1+|\alpha_1|/m} \left[\text{ad}_z^{\alpha_1} \text{ad}_{D_z}^{\beta_1}(A - \lambda)^{-1} \right] \in \mathcal{L}(L^2) \text{ for } (\alpha_1,\beta_1) < (\alpha,\beta) .$$

Here $(\alpha_1,\beta_1) < (\alpha,\beta)$ means $\{\alpha_1 \le \alpha, \beta_1 \le \beta, \text{ and } \alpha_1 + \beta_1 < \alpha + \beta\}$. When $\lambda = -t$ the uniform estimates of the seminorms of $(A+t)^{-1} \in \text{Op } S_\Psi^{-m}$ are a consequence of

$$\|A(A + t)^{-1}\| \le 1 .$$

∎

4.4 Powers of Positive Elliptic Operators

We conclude this chapter with a result about powers of positive elliptic operators.

Theorem 4.8.
Let $A \in \text{Op } S_\Psi^m$, $m \ge 1$, be a positive operator, $A \ge c_0 \text{Id}$, $c_0 > 0$, with

$A = a^W(z, D_z)$, $a \geq C^{-1}\Psi^m - R$, $R \in S_\Psi^{m-\delta}$, $\delta > 0$. Then for any real $s \in \mathbb{R}$, A^s belongs to $\operatorname{Op} S_\Psi^{ms}$ and there exists a constant $t_0 > 0$ such that for all $s \in \mathbb{R}$

$$A^s - [(t_0 + a)^s]^W \in \operatorname{Op} S_\Psi^{ms-1} . \tag{4.18}$$

For a constant $t_0 > 0$ large enough, the ellipticity assumption implies that $t_0 + A = t_0 + a^W(z, D_z)$ is fully elliptic:

$$t_0 + a \geq C_1^{-1}\Psi^m, \quad C_1 > 0 .$$

The identity (4.17) gives for any $t \geq 0$

$$(t + t_0 + A) \circ [(t + t_0 + a)^{-1}]^W = 1 + B_{2,t} ,$$

where the seminorms of $B_{2,t}$ in $\operatorname{Op} S_\Psi^{-1}$ are bilinearly controlled by the seminorms of $A \in \operatorname{Op} S_\Psi^m$ and the seminorms of $(t + t_0 + a)^{-1} \in S_\Psi^{-m+1}$. These seminorms of $(t + t_0 + a)^{-1}$ in S_Ψ^{-m+1} are of order $O(\langle t \rangle^{-1/m})$. Moreover the seminorms of $(t + t_0 + A)^{-1} \in \operatorname{Op} S_\Psi^{-m}$ are uniformly bounded according to Proposition 4.7. Hence we get

$$\forall t \geq 0, (t + t_0 + A)^{-1} - [(t + t_0 + a)^{-1}]^W = E_t \in \operatorname{Op} S_\Psi^{-m-1} ,$$

with all the seminorms of E_t in $\operatorname{Op} S_\Psi^{-m-1}$ bounded by $\langle t \rangle^{-1/m}$.

The first resolvent identity gives, for any $t \geq 0$,

$$(t + A)^{-1} = (t + t_0 + A)^{-1} + t_0(t + A)^{-1}(t + t_0 + A)^{-1} .$$

With $t = 0$, this leads to

$$A^{-1} = [(t_0 + a)^{-1}]^W + E_0 + t_0 A^{-1}(t_0 + A)^{-1} .$$

Since A^{-1} and $(t_0 + A)^{-1}$ belong to $\operatorname{Op} S_\Psi^{-m}$ with $m \geq 1$ and $E_0 \in \operatorname{Op} S_\Psi^{-m-1}$, the pseudo-differential calculus gives

$$\forall p \in \mathbb{Z}, A^p - [(t_0 + a)^p]^W \in \operatorname{Op} S_\Psi^{mp-1} .$$

Let us consider now the case of non integer exponents $r \in \mathbb{R}$. Indeed the pseudo-differential calculus reduces the problem to

$$\forall r \in I, A^r - [(t_0 + a)^r]^W \in \operatorname{Op} S_\Psi^{mr-1} ,$$

for some open interval I of $\mathbb{R}$. We will use the formula (see for example [Yos])

$$A^r = -\frac{\sin(\pi r)}{\pi} \int_0^\infty dt \, t^r (t + A)^{-1}, \quad r \in (-1, 0) . \tag{4.19}$$

We iterate the first resolvent formula:

$$(t + A)^{-1} = (t + t_0 + A)^{-1} + t_0(t + t_0 + A)^{-2} + t_0^2(t + A)^{-1}(t + t_0 + A)^{-2}$$
$$= (1) + (2) + (3) .$$

Term (1):

The identity

$$(t + t_0 + A)^{-1} = [(t + t_0 + a)^{-1}]^W + E_t \tag{4.20}$$

will be used as it is, but it also implies that all the seminorms of $(t + t_0 + A)^{-1}$ in $\mathrm{Op}\, S^{-m+1}$ are bounded by $\langle t \rangle^{-1/m}$.

Term (2):

The equality (4.10) gives

$$(t + t_0 + a)^{-1} \#^W (t + t_0 + a)^{-1} = (t + t_0 + a)^{-2} + R_2((t + t_0 + a)^{-1}(t + t_0 + a)^{-1}),$$

where the seminorms of $R_2((t + t_0 + a)^{-1}, (t + t_0 + a)^{-1})$ in S_Ψ^{-2m} and therefore in S_Ψ^{-m-1} are controlled by the seminorm of $(t_0 + t + a)^{-1}$ in S_Ψ^{-m+1}. With (4.20), this yields

$$(t + t_0 + A)^{-2} = [(t + t_0 + a)^{-2}]^W + E'(t), \tag{4.21}$$

with all the seminorms of $E'(t)$ in $\mathrm{Op}\, S^{-m-1}$ bounded by $\langle t \rangle^{-1/m}$.

Term (3):

The operator $(t + A)^{-1}$ and $(t + t_0 + A)^{-1}$ are uniformly bounded in $\mathrm{Op}\, S_\Psi^{-m}$ while all the seminorms of $(t + t_0 + A)^{-1}$ in $\mathrm{Op}\, S_\Psi^{-m+1}$ are bounded by $\langle t \rangle^{-1/m}$. With $m \geq 1$ again, the seminorms of the term (3) in $\mathrm{Op}\, S_\Psi^{-m-1}$ are bounded by $\langle t \rangle^{-1/m}$.

The three above estimates lead to

$$(t + A)^{-1} = [(t + t_0 + a)^{-1}]^W + [(t + t_0 + a)^{-2}]^W + E''(t),$$

with all the seminorms of $E''(t)$ in $\mathrm{Op}\, S_\Psi^{-m-1}$ bounded by $\langle t \rangle^{-1/m}$. With $r \in (-1, -1 + 1/m)$, the integral

$$\int_0^{+\infty} dt \, t^r E''(t)$$

converges in $\mathrm{Op}\, S_\Psi^{-m-1}$, while we have

$$-\frac{\sin(\pi r)}{\pi} \int_0^\infty dt \, t^r (t + t_0 + a)^{-1} = (t_0 + a)^r \tag{4.22}$$

and

$$-\frac{\sin(\pi r)}{\pi} \int_0^\infty dt \, t^r (t + t_0 + a)^{-1} = -r(t_0 + a)^{r-1}. \tag{4.23}$$

We have proved

$$(t + A)^r = [(t_0 + a)^r]^W - r t_0 [(t_0 + a)^{r-1}]^W + F_r, \quad \text{for } F_r \in \mathrm{Op}\, S_\Psi^{-m-1},$$
$$= [(t_0 + a)^r]^W + F'_r, \quad \text{for } F'_r \in \mathrm{Op}\, S_\Psi^{rm-1},$$

and for all $r \in I = (-1, -1 + 1/m)$.   ∎

4.5 Comments

i) The condition $m \geq 1$ is necessary for $-t_0 = A - (t_0 + a)^W$ cannot belong to $\operatorname{Op} S_\Psi^{m-1}$ if $m < 1$.

ii) Several formulas for the functional calculus are available in order to prove that powers of pseudo-differential operators are pseudo-differential operators, or more generally in order to estimate commutators with powers of some self-adjoint or sectorial operators. For sectorial operators A, one can use contour integrals involving the exponential (see [Yos]). For self-adjoint operators which are not semi-bounded (like the Dirac operator), one can use the Dynkin-Helffer-Sjöstrand formula (see [HelSj6], [Ni2] or Davies [Dav2]) for defining the functional calculus and recognize the pseudo-differential character of the function of the operator. For a positive operator any version of (4.19) works.

iii) The analysis of powers of operators in connection with the pseudo-differential operators now has a long history. It goes back at least to the work of R. Seeley in [See] for operators on a compact manifold and was generalized in the same spirit for operators on $\mathbb{R}^n$ by D. Robert in [Rob]. Let us explain the difference with their strategy and their result. First notice that our result is valid with $\Psi = 1$ (then Theorem 4.8 says nothing but $A^s \in \operatorname{Op} S_\Psi^0$ because there is no notion of principal symbol) and all intermediate cases where Ψ grows partially to ∞. In [See] and [Rob] there is a clear asymptotics with notions of principal symbols and asymptotic expansion up to operators of order $-\infty$. It is not the present case. More generally there are two strategies to attack the pseudo-differential properties of an operator provided by functional analysis:

1. If the pseudo-differential calculus contains a clear asymptotics (high frequencies, semiclassical asymptotics, spatial weights) then one can follow the usual approach by studying recursively all the terms of some asymptotic expansion;

2. Use the Beals criterion in the framework of global pseudo-differential calculus. This criterion introduced by Beals in [Be1][Be2][Be3] was clarified by Bony and Chemin in [BonChe] after the introduction of the biconfinement inequalities in [BoLe].

iv) In [Bon1], J.M. Bony gave another version of the Beals criterion under the assumption $g^\sigma = \lambda^2 g$ and g **geodesically temperate** . Under this additional assumption and with this new criterion, it is possible to prove that the inverse of an invertible operator in $\mathcal{L}(L^2(\mathbb{R}^d))$ is a pseudo-differential operator, without any ellipticity assumption. With the help of Dynkin-Helffer-Sjöstrand formula [Dav1] it is then possible to study the functions of self-adjoint pseudo-differential operators with this more direct version of the Beals criterion. For general metrics, it is not known whether the temperance implies the geodesic temperance (the other implication is true). We shall come back to this point in Section 4.7 where we reproduce a remark communicated to us by J.M. Bony after reading the first version of this text.

v) The analysis of globally elliptic or globally hypoelliptic operators has also a long story. In addition to the previous references, let us also mention Tulovskii-Shubin [TuSh], Robert [Rob], Helffer [Hel0], Mohamed [Moh] and more recently [Glo3] and [BoBuRo] (and references therein).

vi) The global pseudo-differential calculus was recently used for Witten Laplacians in [Iwas] in connection with the Hodge-Kodaira theory.

4.6 Other Types of Pseudo-differential Calculus

For conciseness and clarity, we did not present the pseudo-differential calculus in its full generality. However the general theory applies in many different situations. Often the main difficulty for a specific application is reduced to finding the right metric. Many works have already been done in this spirit and when one studies compactness properties of resolvent, the basic example is associated with the metric:

$$(1 + |z|^2 + |\zeta|^2)^{-1}(dz^2 + d\zeta^2) .$$

We refer the reader to [Hel0] for a detailed presentation of this calculus. There the class of symbols, which satisfy

$$\left| \partial_{z,\zeta}^\alpha a(z,\zeta) \right| \leq C_\alpha \, \langle z, \zeta \rangle^{m - \rho |\alpha|} ,$$

was denoted by Γ_ρ^m, $0 < \rho \leq 1$, and the Sobolev scale associated to this calculus is the one given by the harmonic oscillator $\Delta_z + |z|^2$. Unfortunately, this classical example cannot be applied to the analysis of the Fokker-Planck operator, except in the case when the potential is (almost) quadratic. We will use it in a paragraph devoted to the quadratic case and we will need the additional standard comparison with the anti-Wick calculus (sometimes called Wick-calculus by some authors, see [Shu], [BeSh], [Hel0] and [Ler]).

As discussed in Shubin [Shu] or in [Hel0], one gets the same class of operators when using the Weyl calculus, the standard calculus or the anti-Wick calculus. The anti-Wick quantization is the positive quantization modelled on Gaussians and given by

$$a^{A-Wick}(z, D_z) := \int_{\mathbb{R}^{2d}} a(\tau) \Pi_\tau \frac{d\tau}{(2\pi)^d} ,$$

where Π_τ is the projector associated to the Gaussian $\phi_\tau \in L^2(\mathbb{R}^d)$

$$\phi_\tau(z) = (\pi)^{-d/4} e^{iz\tau_2/2} e^{-\frac{1}{2}|z-\tau_1|^2} .$$

It is immediate to see that $a^{A-Wick}(z, D_z)$ is positive if a is positive and one can verify that

$$a^{A-Wick}(z, D_z) = \left(a * (\pi)^{-d} e^{-|z|^2 - |\zeta|^2}\right)^{Weyl}(z, D_z).$$

A consequence is that[4] for any $a \in \Gamma_1^m$ the difference

$$a^{Weyl}(z, D_z) - a^{A-Wick}(z, D_z) \in \mathrm{Op}\,\Gamma_1^{m-2} . \tag{4.24}$$

In particular, if $m \geq 2$, this difference is bounded in $\mathcal{L}(L^2(\mathbb{R}^d))$. Indeed the general comparison between anti-Wick and Weyl quantization is accessible for many metrics including metrics of the type $dz^2 + \frac{d\zeta^2}{\psi^2}$ by following a process of partition of unity which relies on the slowness and temperance properties (see for example [Ler]) but we will not need it in the sequel. Note that the remainder of (4.24) is in $\mathrm{Op}\,\Gamma_1^{m-2}$ because the gain associated with this calculus is:

$$\lambda(Z) = \left(\min_{T \neq 0} \frac{g^\sigma(T)}{g(T)}\right)^{1/2} = (1 + |z|^2 + |\zeta|^2) .$$

4.7 A Remark by J.M. Bony
About the Geodesic Temperance

In [Bon1] and [Bon2], J.M. Bony introduced the geodesic temperance condition which looks stronger than the temperance:

$$\left(\frac{g_Z}{g_{Z'}}\right)^{\pm 1} \leq C\,(1 + \mathrm{dist}_{g^\sigma}(Z, Z'))^N ,$$

where dist_{g^σ} is the geodesic distance associated with the dual metric g^σ. For the metrics which satisfy this condition (with $g^\sigma = \lambda^2 g$, as is the case for us) there is a good theory of Fourier integral operators and the Beals criterion can be simplified. For this last point, one considers the class S_g^+ of C^∞ functions ℓ on $\mathbb{R}^{2d}$ which satisfy

$$|T_0 T_1 \ldots T_N\, \ell(Z)| \leq C_{N,\ell} ,$$

for any finite family, $N \in \mathbb{N}$, of vector fields such that $g(T_i) \leq 1$ for $1 \leq i \leq N$ and $g^\sigma(T_0) \leq 1$ hold uniformly.
An operator $A : \mathcal{S}(\mathbb{R}^d) \to \mathcal{S}'(\mathbb{R}^d)$ belongs to $\mathrm{Op}\,S(1, g)$ if and only if for any finite family $\ell_1, \ldots, \ell_N$ in S_g^+, $N \in \mathbb{N}$, one has

$$\mathrm{ad}_{L_1} \ldots \mathrm{ad}_{L_N} A \in \mathcal{L}(L^2(\mathbb{R}^d)) \qquad L_j = \ell_j^W(z, D_z) .$$

Then the inverse of any invertible $A \in \mathrm{Op}\,S(1, g)$ belongs to $\mathrm{Op}\,S(1, g)$ and the use of this Beals criterion can simplify the analysis of functions of self-adjoint pseudodifferential operators.

---

[4] The proof can for example be found in [Hel0] (Chapitre 1, Theorem 1.4.6)

After reading the initial version of this text, J.-M. Bony [Bon3] showed us that for the metrics of the form $g = dz^2 + \frac{d\zeta^2}{\psi(Z)}$ in which we are interested, the temperance implies the geodesic temperance. We reproduce here his proof.

Theorem 4.9.
Let g be a metric on $\mathbb{R}^{2d}_{z,\zeta} = \mathbb{R}^{2d}_Z$ of the form

$$g = dz^2 + \frac{d\zeta^2}{\Psi(Z)^2} \ , \quad with \ \Psi \geq 1 \ ,$$

and which satisfies Hörmander's slowness and temperance conditions. Then g is geodesically temperate.

Lemma 4.10.
Let γ be a metric in $\mathbb{R}^\nu$ such that for any $x, y \in \mathbb{R}^\nu$, γ_x and γ_y are proportional with the uniform estimate

$$\left(\frac{\gamma_x}{\gamma_y} \right)^{\pm 1} \leq C \left(1 + \gamma_x(x-y) \right)^N \ .$$

Then there exist two positive constants $C' > 0$ and $N' > 0$ such that

$$\forall x, y \in \mathbb{R}^\nu, \ \gamma_x(x-y) \leq C' \left(1 + \operatorname{dist}_\gamma(x,y) \right)^{N'} \ .$$

Let $t \to x_t$ be a C^1 curve going from x to y with γ-length ℓ and set $R = \gamma_x(x-y)^{1/2}$. It suffices to consider the case $R \geq 1$. Let t_1 be the largest time such that

$$\gamma_x(x_t - x)^{1/2} \leq R/2$$

and let t_2 be the smallest $t > t_1$ such that

$$\gamma_x(x_t - x)^{1/2} \geq R \ .$$

The curve $t \to x_t$ is parametrized such that $\gamma_{x_t}(\dot{x}_t) = \gamma_{x_t}(x_t - x)$. Then the proportionality assumption yields $\gamma_x(\dot{x}_t) = \gamma_x(x_t - x)$, where this quantity has to be smaller than R^2 in the interval $[t_1, t_2]$. The Cauchy-Schwarz inequality and the triangular inequality for γ_x then lead to:

$$\frac{3}{4}R^2 = \gamma_x(x_{t_2} - x) - \gamma_x(x_{t_1} - x)$$

$$\leq \gamma_x(x_{t_2} - x_{t_1})^{1/2} \gamma_x \left(x_{t_2} - x + x_{t_1} - x \right)^{1/2}$$

$$\leq \int_{t_1}^{t_2} \gamma_x(\dot{x}_t)^{1/2} \ dt \times \frac{3R}{2}$$

$$\leq \frac{3R^2}{2}(t_2 - t_1) \ .$$

We must have $t_2 - t_1 \geq 1/2$. Finally we use the temperance inequality with $\gamma_x(x_t - x) \geq 1/4$ for $t \in [t_1, t_2]$ $(R \geq 1)$, in

$$\ell \geq \int_{t_1}^{t_2} \gamma_{x_t}(\dot{x}_t) \, dt \geq \int_{t_1}^{t_2} \gamma_{x_t}(x_t - x)^{1/2} \, dt$$

$$\geq C_0^{-1} \int_{t_1}^{t_2} \gamma_x(x_t - x)^{1/(2N+2)} \, dt$$

$$\geq C_1^{-1} R^{1/(N+1)} \, .$$

∎

Proof of Theorem 4.9:
The dual metric equals here $g^\sigma = \Psi(Z)^2 dz^2 + d\zeta^2$. For $X, Y \in \mathbb{R}^{2d}$ we call $d(X, Y)$ their geodesic distance for g^σ. The small parameter $\varepsilon > 0$ will be fixed in the end. It suffices to prove

$$\forall X, Y \in \mathbb{R}^{2d}, \ g_X^\sigma(X - Y) \leq C'(1 + d(X, Y))^{N'}.$$

We recall $X = (x, \xi)$ and $Y = (y, \eta)$ and

$$T_{XY} = \left\{ Z = (z, \zeta) \in \mathbb{R}^{2d}, \ |\zeta - \xi| < g_X^\sigma(X - Y)^\varepsilon \right\} \, .$$

There are two cases:
a) A g^σ-geodesic curve going from X to Y leaves the region T_{XY}. Then there exists $Z_0 \notin T_{XY}$ such that

$$d(X, Y) \geq d(X, Z_0) \geq |\zeta_0 - \xi| \geq g_X^\sigma(X - Y)^\varepsilon \, .$$

b) A g^σ-geodesic curve $t \to Z_t$ between $X = Z_0$ and $Y = Z_1$ lies in T_{XY}. We set $Z_t' = (z_t, \xi)$ and $Y' = Z_1' = (y, \xi)$. In this case we have

$$g_{Z_t}^\sigma(Z_t - Z_t') = |\zeta_t - \xi|^2 \leq g_X^\sigma(X - Y)^{2\varepsilon} \, .$$

The temperance property then gives

$$\left(\frac{g_{Z_t}}{g_{Z_t'}} \right)^{\pm 1} \leq C \left(1 + g_X^\sigma(X - Y) \right)^{2N\varepsilon} \, .$$

Denoting by $\delta(x, y)$ the geodesic distance for the metric $\Psi(z, \xi)^2 dz^2$, the previous inequality yields

$$d(X, Y) \geq C^{-1} \left(1 + g_X^\sigma(X - Y) \right)^{-N\varepsilon} \delta(x, y)$$

and

$$1 + d(X, Y) \geq C^{-1} \left(1 + g_X^\sigma(X - Y) \right)^{-N\varepsilon} (1 + \delta(x, y)) \, .$$

The temperance condition on g ensures that the metric $\psi(z,\xi)dz^2$ in $\mathbb{R}_z^d$ satisfies the assumptions of Lemma 4.10 with constants C, N independent of $\xi \in \mathbb{R}^d$. Hence there are some constants $C_1 > 0$ and $N_1 > 0$ such that

$$1 + \delta(x,y) \geq C_1^{-1}(1 + g_X^\sigma(Y' - X))^{1/N_1} .$$

With

$$g_X^\sigma(Y - X) \leq 2g_X^\sigma(Y' - X) + 2|\eta - \xi|^2 \leq 2g_X^\sigma(Y' - X) + 2g_X^\sigma(Y - X)^\varepsilon ,$$

we obtain

$$1 + \delta(x,y) \geq C_2^{-1}(1 + g_X^\sigma(Y - X))^{1/N_1} ,$$

and therefore

$$1 + d(X,Y) \geq C_3^{-1}(1 + g_X^\sigma(X - Y))^{1/(N_1 - N\varepsilon)} .$$

We conclude by taking $\varepsilon \leq 1/(2N_1 N)$. ∎

5

Analysis of Some Fokker-Planck Operator

5.1 Introduction

The analysis presented in this chapter is an application of the Kohn method. The same ingredients are indeed present: existence of a family of Λ^s and commutator techniques. We follow with some improvements the analysis given by Hérau-Nier in [HerNi]. For the Fokker-Planck equation it seems more efficient to work with the creation-annihilation operators $-\partial_v + \frac{v}{2}$, $\partial_v + \frac{v}{2}$, of the harmonic oscillator and their distorted version associated with the Witten Laplacian $-\partial_x + \frac{1}{2}\partial_x V$ and $\partial_x + \frac{1}{2}\partial_x V$, rather than separating ∂_v, v, ∂_x and $\partial_x V(x)$. Nevertheless the work of J.P. Eckmann and M. Hairer relies in a more general context on this second approach and we refer the reader to [EckHai1, EckHai2] for details in this direction.

Following [HerNi], we would like to analyze the links between the compact resolvent property for the Fokker-Planck operator and the same property for the corresponding Witten Laplacian.

5.2 Maximal Accretivity of the Fokker-Planck Operator

5.2.1 Accretive Operators

We collect here some material on accretive operators. The references could be the books by Dautray-Lions (Vol. 5, Chapter XVII), Reed-Simon [ReSi] or [Dav1]. Let $\mathcal{H}$ be a complex (or real) Hilbert space.

Definition 5.1.
Let A be an unbounded operator in $\mathcal{H}$ with domain $D(A)$. We say that A is accretive if

$$\operatorname{Re} \langle Ax \mid x \rangle_{\mathcal{H}} \geq 0 \ , \ \forall x \in D(A) \ . \tag{5.1}$$

Definition 5.2.
An accretive operator A is maximally accretive if it does not exist an accretive extension $\tilde{A}$ with strict inclusion of $D(A)$ in $D(\tilde{A})$.

Proposition 5.3.
Let A be an accretive operator with domain $D(A)$ dense in $\mathcal{H}$. Then A is closable and its closed extension $\overline{A}$ is accretive.

For the analysis of the Fokker-Planck operator, the following criterion, which extends the standard criterion of essential self-adjointness, will be the most suitable

Theorem 5.4.
For an accretive operator A, the following conditions are equivalent

1. *$\overline{A}$ is maximally accretive.*
2. *There exists $\lambda_0 > 0$ such that $A^* + \lambda_0 I$ is injective.*
3. *There exists $\lambda_1 > 0$ such that the range of $A + \lambda_1 I$ is dense in $\mathcal{H}$.*

Note that in this case $-\overline{A}$ is the infinitesimal generator of a contraction semi-group.

5.2.2 Application to the Fokker-Planck Operator

We would like to show

Proposition 5.5.
Let V be a C^∞ potential on $\mathbb{R}^n$, then the closure $\overline{K}$ of the Fokker-Planck operator defined on $C_0^\infty(\mathbb{R}^{2n})$ by

$$K := -\Delta_v + \frac{1}{4}|v|^2 - \frac{n}{2} + X_0 , \tag{5.2}$$

where

$$X_0 := -\nabla V(x) \cdot \partial_v + v \cdot \partial_x \tag{5.3}$$

is maximally accretive.
Moreover K^* is also maximally accretive.

The idea is to adapt the proof that a semi-bounded Schrödinger operator with regular potential is essentially self-adjoint on $L^2(\mathbb{R}^n)$ (See for example Theorem 6.6.2 in [Hel11] for a proof of this result of Simader). The result is already known under additional restrictions, see Hérau–Nier [HerNi], Eckmann–Pillet–Rey-Bellet [EckPiRe-Be] Eckmann–Hairer [EckHai1, EckHai2], Rey-Bellet–Thomas [Re-BeTh1, Re-BeTh2, Re-BeTh3].

Proof:
We apply the abstract criterion taking $\mathcal{H} = L^2(\mathbb{R}^n \times \mathbb{R}^n)$ and $A = K$. The operators being real, we can consider everywhere real functions. The accretivity on $C_0^\infty(\mathbb{R}^{2n})$ is clear. We can then consider the closure $\overline{K}$.
    Changing K in $T := K + (\frac{n}{2} + 1)I$, we would like to show that its range is dense.
Let $f \in L^2(\mathbb{R}^m)$, with $m = 2n$, such that

$$< f \mid Tu >_{\mathcal{H}} = 0 \,, \ \forall u \in C_0^\infty(\mathbb{R}^m) \,. \tag{5.4}$$

We have to show that $f = 0$.
Because K is real, one can assume that f is real.
We first observe that (5.4) implies that:

$$(-\Delta_v + v^2/4 + 1 - X_0)f = 0 \,, \ \text{in} \ \mathcal{D}'(\mathbb{R}^m) \,.$$

The standard hypoellipticity theorem for the Hörmander operators of type 2
(See Section 2.2) implies that $f \in C^\infty(\mathbb{R}^m)$.
We now introduce a family of cut-off functions $\zeta_k := \zeta_{k_1,k_2}$ by

$$\zeta_{k_1,k_2}(x,v) := \zeta(x/k_1)\zeta(v/k_2) \,, \ \forall k \in \mathbb{N}^2 \,, \tag{5.5}$$

where ζ is a C^∞ function satisfying $0 \le \zeta \le 1$, $\zeta = 1$ on $B(0,1)$ and $\mathrm{supp}\,\zeta \subset B(0,2)$.
For any $u \in C_0^\infty$, we have the identity

$$\begin{aligned}
&\int \nabla_v(\zeta_k f) \cdot \nabla_v(\zeta_k u) \,dxdv + \int \zeta_k(x,v)^2(v^2/4+1)u(x,v)\,f(x,v)\,dx\,dv \\
&+ \int f(x,v)(X_0(\zeta_k^2 u))(x,v)\,dx\,dv \\
&= \int |(\nabla_v\zeta_k)(x,v)|^2 u(x,v)f(x,v)\,dx\,dv \\
&\quad + \sum_{i=1}^m \int (f(\partial_{v_i}u) - u(\partial_{v_i}f))\,(x,v)\zeta_k(x,v)(\partial_{v_i}\zeta_k)(x,v)\,dx\,dv \\
&\quad + \langle f(x,v) \mid T\zeta_k^2 u \rangle \,.
\end{aligned} \tag{5.6}$$

When f satisfies (5.4), we get:

$$\begin{aligned}
&\int_{\mathbb{R}^m} \nabla_v(\zeta_k f) \cdot \nabla_v(\zeta_k u) \,dxdv + \int \zeta_k^2(v^2/4+1)u(x,v)\,f(x,v)\,dx\,dv \\
&+ \int f(x,v)(X_0(\zeta_k^2 u))(x,v)dx\,dv \\
&= \int |(\nabla_y\zeta_k)(x)|^2 u(x)f(x,v)\,dx\,dv \\
&\quad + \sum_{i=1}^m \int (f(\partial_{v_i}u) - u(\partial_{v_i}f))\,(x,v)\zeta_k(x,v)(\partial_{v_i}\zeta_k)(x,v)\,dx\,dv \,,
\end{aligned} \tag{5.7}$$

for all $u \in C^\infty(\mathbb{R}^m)$. In particular, we can take $u = f$.
We obtain

$$\begin{aligned}
&< \nabla_v(\zeta_k f) \mid \nabla_v(\zeta_k f) > + \int \zeta_k^2(v^2/4+1)|f(x,v)|^2 \,dx\,dv \\
&+ \int f(x,v)(X_0(\zeta_k^2 f))(x,v)\,dx\,dv \\
&= \int |\nabla_v\zeta_k|^2|f(x,v)|^2 \,dx\,dv \,.
\end{aligned} \tag{5.8}$$

With an additional integration by part, we get

$$\begin{aligned}
&< \nabla_v(\zeta_k f) \mid \nabla_v(\zeta_k f) > + \int \zeta_k^2(v^2/4+1)|f(x,v)|^2 \,dx\,dv \\
&+ \int \zeta_k f(x,v)^2(X_0\zeta_k)(x,v)\,dx\,dv \\
&= \int |\nabla_v\zeta_k|^2|f(x,v)|^2 \,dx\,dv \,.
\end{aligned} \tag{5.9}$$

This leads to the existence of a constant C such that, for all k,

$$\begin{aligned}
&\|\zeta_k f\|^2 + \tfrac{1}{4}\|\zeta_k v f\|^2 \\
&\le C\tfrac{1}{k_2^2}\|f\|^2 + C\tfrac{1}{k_1}\|v\zeta_k f\|\,\|f\| + C\tfrac{1}{k_2}\|\nabla V(x)\zeta_k f\|\,\|f\| \,.
\end{aligned} \tag{5.10}$$

(The constant C will possibly be changed from line to line). This leads to

$$||\zeta_k f||^2 + \frac{1}{8}||\zeta_k v f||^2 \leq C(\frac{1}{k_2^2} + \frac{1}{k_1^2})||f||^2 + C(k_1)\frac{1}{k_2}||\zeta_k f|| \, ||f|| \,, \quad (5.11)$$

where

$$C(k_1) = \sup_{|x| \leq 2k_1} |\nabla_x V(x)|$$

This implies

$$||\zeta_k f||^2 \leq C(\frac{\tilde{C}(k_1)}{k_2^2} + \frac{1}{k_1^2})||f||^2 \,. \quad (5.12)$$

This finally leads to $f = 0$. For example, one can take first the limit $k_2 \to +\infty$, which leads to

$$||\zeta(\frac{x}{k_1})f||^2 \leq \frac{C}{k_1^2}||f||^2 \,,$$

and then the limit $k_1 \to +\infty$.

5.3 Sufficient Conditions for the Compactness of the Resolvent of the Fokker-Planck Operator

5.3.1 Main Result

Let us introduce some convenient notations. We observe that the operator K defined in (5.2) and (5.3)) can be written

$$K = X_0 + b^* b \,, \quad (5.13)$$

where

$$X_0 = (b^* a - a^* b) \,. \quad (5.14)$$

with

$$b = \partial_v + \frac{v}{2} = \begin{pmatrix} b_1 \\ \vdots \\ b_n \end{pmatrix} \,, \quad a = \partial_x + \frac{1}{2}\partial_x V = \begin{pmatrix} a_1 \\ \vdots \\ a_n \end{pmatrix} \,. \quad (5.15)$$

The adjoint forms of a and b are

$$b^* = (b_1^*, \ldots, b_n^*) \quad \text{and} \quad a^* = (a_1^*, \ldots, a_n^*); \,. \quad (5.16)$$

With these notations we introduce the operator Λ defined by

$$\Lambda^2 = 1 + a^* a + b^* b \,,$$

and which provides a natural Sobolev scale for the problem. Note that the operator

$$\Lambda^2 - 1 = a^* a + b^* b = \Delta^{(0)}_{\Phi/2} = \Delta^{(0)}_{V/2} \otimes \mathrm{Id}_v + \mathrm{Id}_x \otimes \Delta^{(0)}_{v^2/4}$$

is the phase-space Witten Laplacian associated to $\Phi/2$, with $\Phi = \frac{v^2}{2} + V(x)$. In order to establish the main theorem, let us introduce some notations and assumptions. We first introduce the notation

$$h(x) = \sqrt{1 + |\nabla V(x)|^2} \, ,$$

and

Assumption 5.6.
The potential $V(x)$ belongs to $C^\infty(\mathbb{R}^n)$ and satisfies:

$$\forall \alpha \in \mathbb{N}^n \, , |\alpha| \geq 1, \forall x \in \mathbb{R}^n \quad |\partial_x^\alpha V(x)| \leq C_\alpha h(x) \, , \tag{5.17}$$

$$\exists M, C \geq 1, \forall x \in \mathbb{R}^n, \quad h(x) \leq C \langle x \rangle^M \, , \tag{5.18}$$

and the coercivity condition

$$\exists M, C \geq 1, \forall x \in \mathbb{R}^n, \quad C^{-1} \langle x \rangle^{1/M} \leq h(x) \, . \tag{5.19}$$

Assumption 5.7.
The potential $V(x)$ belongs to $C^\infty(\mathbb{R}^n)$ and satisfies (5.17) (5.18) with the coercivity condition (5.19) replaced by the existence of $\rho_0 > 0$ and $C > 0$ such that:

$$\forall x \in \mathbb{R}^n, \quad |\nabla h(x)| \leq C \, h(x) \langle x \rangle^{-\rho_0} \, . \tag{5.20}$$

Theorem 5.8.
If the potential $V \in C^\infty(\mathbb{R}^n)$ verifies Assumption 5.6 or Assumption 5.7, then there exists a constant $C > 0$ such that

$$\forall u \in \mathcal{S}(\mathbb{R}^{2n}), \, \left\| \Lambda^{1/4} u \right\|^2 \leq C \left(\| K u \|^2 + \| u \|^2 \right) \, . \tag{5.21}$$

Remark 5.9.
As for the Kohn's proof for the hypoellipticity, the exponent $\frac{1}{4}$ in (5.21) is not optimal. We shall give better results, in the quadratic case, in Subsection 5.5.2 and in Section 9.2.

Corollary 5.10.
If the potential $V \in C^\infty(\mathbb{R}^n)$ satisfies Assumption 5.6 then the operator K has a compact resolvent.
If the potential $V \in C^\infty(\mathbb{R}^n)$ satisfies Assumption 5.7, then K has a compact resolvent if (and only if) the Witten Laplacian $\Delta^{(0)}_{V/2}$ has a compact resolvent.

Proof:
The closure of K, initially defined on $\mathcal{S}(\mathbb{R}^n)$, is maximally accretive according to Proposition 5.5. Theorem 5.8 says that the first factor of

$$(1+K)^{-1} = \left[(1+K)^{-1}\Lambda^{1/4}\right]\Lambda^{-1/4}$$

is bounded, while the second one belongs to the class $\mathrm{Op}\,S_\Psi^{-1/4}$ which is specified below. Under Assumption 5.6, the function Ψ satisfies

$$\lim_{(x,v,\xi,\eta)\to\infty} \Psi(x,v,\xi,\eta) = +\infty$$

and $\Lambda^{-1/4}$ is compact. This last condition is not implied by Assumption 5.7 but the compactness of $\Lambda^{-1/4}$ is then a consequence of the compactness of $\Lambda^{-2} = \left(1+\Delta_{V/2}^{(0)}+\Delta_{v^2/2}^{(0)}\right)^{-1}$. The "only if" part will be discussed in Section 5.4. ∎

5.3.2 A Metric Adapted to the Fokker-Planck Equation and Weak Ellipticity Assumptions

We will apply the results of Chapter 4 with a metric adapted to the analysis of the Fokker-Planck equation and more precisely to the resolvent of the associated Witten Laplacian

$$a^*a + b^*b = -\Delta_x + \frac{1}{4}|\nabla_x V(x)|^2 - \frac{1}{2}\Delta_x V(x) - \Delta_v + \frac{v^2}{4} - \frac{n}{2} = \Delta_{\Phi/2}^{(0)},$$

with $\Phi(x,v) = v^2/2 + V(x)$. We will consider on $\mathbb{R}_{x,v,\xi,\eta}^{4n} = T^*\mathbb{R}_{x,v}^{2n}$ the metric

$$g = dx^2 + dv^2 + \frac{d\xi^2 + d\eta^2}{\Psi^2}, \tag{5.22}$$

with

$$\Psi(x,\xi,v,\eta)^2 = 1 + |\xi|^2 + |\eta|^2 + \frac{1}{4}|v|^2 + \frac{1}{4}|\nabla V(x)|^2. \tag{5.23}$$

Some assumptions on the potential V appearing in Theorem 5.8 are actually required in order to enter in the global pseudo-differential calculus presented in Chapter 4.

Proposition 5.11.
Under Assumptions 5.6 or 5.7, the metric $g = dx^2 + dv^2 + \frac{d\xi^2+d\eta^2}{\Psi^2}$ satisfies the slowness and temperance properties (4.7)-(4.8).

First of all note that the Assumptions 5.6 or 5.7 for the potential V imply similar properties for the phase space potential $\Phi(x,v) = v^2/2 + V(x)$ (the constant factors are not important here). It suffices to check the slowness and temperance properties for the metric

$$\gamma = dx^2 + \frac{d\xi^2}{|\xi|^2 + h(x)^2} \ .$$

The result for the metric g is derived similarly after replacing x by (x, v), ξ by (ξ, η) and $|\xi|^2 + h(x)^2$ by Ψ^2.

It is a consequence of the following lemma.

Lemma 5.12.

Under Assumptions 5.6 or 5.7, the function $h(x) = \sqrt{1 + |\nabla V(x)|^2}$ satisfies for some constants $C_0, C_1, N_1, N_2 > 0$ the uniform estimates

$$\left(|x - x'| \le C_0^{-1}\right) \Rightarrow \left(\left(\frac{h(x)}{h(x')}\right)^{\pm 1} \le C_0\right), \tag{5.24}$$

and

$$\left(\frac{h(x)}{h(x')}\right)^{\pm 1} \le C_1 h(x)^{N_1} \langle x - x' \rangle^{N_2} \ . \tag{5.25}$$

In this chapter X and X' will denote the variables (x, ξ) and (x', ξ').

Slowness.

Assume $\gamma_X(X - X') \le \delta^2$ with $0 < \delta \le C_0^{-1}$. It implies

$$|x - x'| \le \delta \quad \text{and} \quad |\xi - \xi'|^2 \le \delta^2 \left(|\xi|^2 + h(x)^2\right) .$$

The inequality

$$|\xi|^2 - 2 |\xi - \xi'|^2 \le 2 |\xi'|^2 \le 4 |\xi|^2 + 4 |\xi - \xi'|^2 .$$

yields

$$|\xi|^2 - 2\delta^2 \left(|\xi|^2 + h(x)^2\right) \le 2 |\xi'|^2 \le 4 |\xi|^2 + 4\delta^2 \left(|\xi|^2 + h(x)^2\right) .$$

Moreover the relation (5.24) gives

$$C_0^{-2} h(x) \le h(x') \le C_0^2 h(x) \ .$$

By taking $2\delta^2$ smaller than $\min(1/2, C_0^{-2}/2)$ we find a constant $C_2 > 0$ such that

$$\left(\gamma_X(X - X') \le \delta^2\right) \Rightarrow \left(\sup_{T \ne 0} \left(\frac{\gamma_X(T)}{\gamma_{X'}(T)}\right)^{\pm 1} \le C_2\right). \tag{5.26}$$

Temperance.

According to (5.26) it suffices to consider the case $\gamma_X(X - X') \geq \delta^2$ with $0 < \delta \leq \sqrt{C_0}$ and δ small enough. There are two cases:

a) Assume $|x - x'|^2 \leq \delta^2/2$ and $|\xi - \xi'|^2 \geq \delta^2(|\xi|^2 + h(x)^2)$. Then we get

$$
\begin{aligned}
|\xi'|^2 + h(x')^2 &\leq 2\,|\xi|^2 + 2\,|\xi - \xi'|^2 + C_0^2 h(x)^2 \\
&\leq \frac{2\max(2, C_0^2)}{\delta^2}\,|\xi - \xi'|^2 \\
&\leq C_3 \gamma^\sigma(X - X') \leq C_3 \left(|\xi|^2 + h(x)^2\right)\left(1 + \gamma_X^\sigma(X - X')\right),
\end{aligned}
$$

where we used $h(x) \geq 1$ in the last inequality. Conversely, the lower bound $h(x') \geq 1$ yields

$$
|\xi|^2 + h(x)^2 \leq \frac{1}{\delta^2}\,|\xi - \xi'|^2 \leq C_3'\left(|\xi'|^2 + h(x')^2\right)\left(1 + \gamma_X^\sigma(X - X')\right).
$$

b) Assume $|x - x'|^2 \geq \delta^2/2$. It implies for any $N > 0$, $|x - x'|^N \geq C_{4,N}^{-1}\langle x - x'\rangle^N \leq C_{4,N}^{-1}$ for some positive constant $C_{4,N} > 0$. We write according to the inequality (5.25)

$$
\begin{aligned}
|\xi'|^2 + h(x')^2 &\leq 2\,|\xi|^2 + 2\,|\xi - \xi'|^2 + C_1^2 h(x)^{2+2N_1}\langle x - x'\rangle^{2N_2} \\
&\leq 2(|\xi|^2 + h(x)^2) + 2\,|\xi - \xi'|^2 + \\
&\qquad C_1^2 C_{4,2N_2}(|\xi|^2 + h(x)^2)^{N_1+1}|x - x'|^{2N_2}.
\end{aligned}
$$

The lower bound $h \geq 1$ implies $|\xi - \xi'|^2 \leq (|\xi|^2 + h(x)^2)|\xi - \xi'|^2$. Thus there exist $C_5, C_6 > 0$ such that

$$
\begin{aligned}
|\xi'|^2 + h(x')^2 &\leq C_5(|\xi|^2 + h(x)^2)\left[1 + (|\xi|^2 + h(x)^2)^{N_1}|x - x'|^{2N_2} + |\xi - \xi'|^2\right] \\
&\leq C_6(|\xi|^2 + h(x)^2)\left[1 + \gamma_X^\sigma(X - X')\right]^{1+N_1+N_2}.
\end{aligned}
$$

By the same process we derive from (5.25) the estimates

$$
\begin{aligned}
|\xi|^2 + h(x)^2 &\leq 2\,|\xi'|^2 + 2\,|\xi - \xi'|^2 + C_1^2 h(x')^2 h(x)^{2N_1}\langle x - x'\rangle^{2N_2} \\
&\leq C_6'(|\xi'|^2 + h(x')^2)\left[1 + \gamma_X^\sigma(X - X')\right]^{1+N_1+N_2}.
\end{aligned}
$$

∎

Proof of Lemma 5.12:

The first estimate (5.24) is an easy consequence of $|\nabla h| \leq Ch$. We focus on (5.25) which requires different proofs according to Assumption 5.6 or Assumption 5.7.

Proof under Assumption 5.6 .

We simply write

$$h(x) \leq h(y) + C \sup_{t \in [0,1]} |\nabla h(tx + (1-t)y)| \langle x - y \rangle$$
$$\leq h(y) + C_2 \sup_{t \in [0,1]} h(tx + (1-t)y) \langle x - y \rangle$$
$$\leq h(y) + C_3 \langle x \rangle^M \langle x - y \rangle^{M+1}$$
$$\leq h(y) + C_4 h(x)^{M^2} \langle x - y \rangle^{M+1}$$
$$\leq h(y)[1 + C_4 h(x)^{M^2} \langle x - y \rangle^{M+1}]$$
$$\leq C_5 h(y) h(x)^{M^2} \langle x - y \rangle^{M+1} .$$

Similarly, one obtains:

$$h(y) \leq h(x) + C_4 h(x)^{M^2} \langle x - y \rangle^{M+1}$$
$$\leq h(x) \left[1 + C_4 h(x)^{M^2} \langle x - y \rangle^{M+1} \right] .$$

Proof under Assumption 5.7.
We cut the space $\mathbb{R}^n_x \times \mathbb{R}^n_{x'}$ in two regions:

$$\{ \langle x - x' \rangle \geq \langle x \rangle^{\varrho_1} \} \text{ and } \{ \langle x - x' \rangle \leq \langle x \rangle^{\varrho_1} \} ,$$

where $\varrho_1 \in (0,1)$ will be fixed later in terms of $\rho_0 > 0$.
For $\langle x - x' \rangle \geq \langle x \rangle^{\varrho_1}$, we write

$$h(x') \leq h(x) + |x' - x| \int_0^1 |\nabla h((1-t)x + tx')| \, dt$$

$$\leq h(x) \left[1 + |x' - x| \int_0^1 \langle (1-t)x + tx' \rangle^M \, dt \right]$$

$$\leq C h(x) \langle x \rangle^M \langle x' - x \rangle^{M+1} \leq C' h(x) \langle x - x' \rangle^{(1+1/\rho_1)M+1}.$$

Conversely, one gets $h(x) \leq C'' h(x') \langle x - x' \rangle^{(1+1/\rho_1)M+1}$ by starting with the
inequality $h(x) \leq h(x') + |x' - x| \int_0^1 |\nabla h(tx + (1-t)x')| \, dt$.
For $\langle x - x' \rangle \leq \langle x \rangle^{\varrho_1}$, we set $\varphi(t) = h((1-t)x + tx')$. Assumption 5.7 gives

$$\frac{|\varphi'|}{\varphi} \leq C |x - x'| \langle (1-t)x + tx' \rangle^{-\rho_0} .$$

With $\rho_0 > 0$, the inequality

$$\forall t \in [0,1], \ \frac{\langle x \rangle^{\rho_0}}{\langle (1-t)x + tx' \rangle^{-\rho_0}} \leq C_{\rho_0} \langle t(x - x') \rangle^{\rho_0} \leq C_{\rho_0} \langle x - x' \rangle^{\rho_0}$$

leads to

$$\frac{|\varphi'|}{\varphi} \leq C_2 \frac{\langle x - x' \rangle^{\rho_0 + 1}}{\langle x \rangle^{\rho_0}} \leq C_2 \langle x \rangle^{\rho_1(\rho_0 + 1) - \rho_0} .$$

By taking $\rho_1 \leq \frac{\rho_0}{\rho_0 + 1}$, we obtain $\frac{|\varphi'|}{\varphi} \leq C_2$, which yields

$$\left(\frac{h(x)}{h(x')} \right)^{\pm 1} = \left(\frac{\varphi(0)}{\varphi(1)} \right)^{\pm 1} \leq C_3 \leq C_3 \langle x - x' \rangle^{(1+1/\rho_1)M+1} .$$

∎

Remarks 5.13.
a) Assumptions 5.6 and 5.7 are weak ellipticity assumptions. They ensure that the Witten Laplacian $\Delta^{(0)}_{\Phi/2} = a^*a + b^*b$ is an elliptic operator in $\operatorname{Op} S^2_\Psi$.
In terms of the potential V, the fact that the weight $h(x) = \sqrt{1 + |\nabla V(x)|^2}$ controls the higher order derivatives according to (5.17) is an ellipticity assumption, which is weaker than the one introduced in [HerNi]. For example, Assumptions 5.6 and 5.7 are satisfied by the potential

$$V(x_1, x_2) = x_1^2 x_2^2 + (x_1^2 + x_2^2)^{3/2}$$

but not by the potential

$$V(x_1, x_2) = x_1^2 x_2^2 + (x_1^2 + x_2^2).$$

If one writes the condition (5.20) of Assumption 5.7 in the more explicit (and slightly stronger) form

$$\forall \alpha \in \mathbb{N}^n, |\alpha| = 2, \forall x \in \mathbb{R}^n, \ |\partial_x^\alpha V(x)| \leq C_\alpha h(x),$$

the two types of assumptions appearing in Assumptions 5.6 and 5.7 lead to

$$|\nabla V(x)|^2 - 2\Delta V(x) \geq C^{-1} |\nabla V(x)|^2, \quad \text{for } |x| \geq C.$$

b) Assumption 5.6 implies that $\Delta^{(0)}_{\Phi/2} = a^*a + b^*b$ has a compact resolvent. This is no more the case under Assumption 5.7. With the coercivity condition (5.19), the function $\Psi(x, v, \xi, \eta)$ satisfies:

$$\lim_{(x,v,\xi,\eta) \to \infty} \Psi(x, v, \xi, \eta) = +\infty.$$

Hence the resolvent $(1 + a^*a + b^*b)^{-1} \in \operatorname{Op} S^{-2}_\Psi$ is compact.
Assumption 5.7 holds for $V(x) = \langle x \rangle$ for which ∇V is bounded. Under Assumption 5.7, the hypoelliptic estimate of Theorem 5.8 holds but one can conclude that the Fokker-Planck operator K has a compact resolvent only by adding the assumption that $(1 + \Delta^{(0)}_{V/2})^{-1}$ is compact.
c) A more general "local" (but non temperate) calculus was developed by N. Dencker [Den] in continuation of [Fei] under the weaker assumption that:

$$|\nabla h(x)| \leq C |h(x)|^{1+\delta},$$

for some $\delta < 1$. This condition appeared in (3.16).

5.3.3 Algebraic Properties of the Fokker-Planck Operator

Before starting the proof of Theorem 5.8 let us recall the algebraic properties associated with the Fokker-Planck operator.

The Canonical Commutation Relations (CCR) of the annihilation-creation operators b_j, b_j^* are satisfied:

$$[b_j, b_k] = [b_j^*, b_k^*] = 0 , \qquad [b_j, b_k^*] = \delta_{jk} . \tag{5.27}$$

More generally with $\partial_{x_j} \partial_{x_k} V = \partial_{x_k} \partial_{x_j} V$, we have:

$$[a_j, a_k] = [a_k, a_j] = 0 , \qquad [a_j, a_k^*] = \partial^2_{x_j x_k} V . \tag{5.28}$$

The a's and b's commute with each other

$$[a_j^\dagger, b_k^\sharp] = 0 , \tag{5.29}$$

where $a^\dagger$ (resp. $b^\sharp$) equals a or a^* (resp. b or b^*).
The a_j's, a_j^*'s are in the Lie algebra generated by the b_j's, b_j^*'s and the vector field X_0:

$$[b_j, X_0] = a_j , \qquad [b_j^*, X_0] = a_j^* . \tag{5.30}$$

Similarly, the b_j's and b_j^*'s can be derived from the a_j's, a_j^*'s and X_0

$$[a_j, X_0] = -\sum_{k=1}^{d} \left(\partial^2_{x_j x_k} V \right) b_k , \qquad [a_j^*, X_0] = -\sum_{k=1}^{d} b_k^* \left(\partial^2_{x_k x_j} V \right) . \tag{5.31}$$

For any $r, r' \in \mathbb{R}$, we have:

$$\left[\Lambda^r, (1 + a^*a)^{r'} \right] = \left[\Lambda^r, (1 + b^*b)^{r'} \right] = 0 . \tag{5.32}$$

The relations (5.30) and (5.31) are summarized by

$$\begin{aligned} [b, X_0] &= a , & [b^*, X_0] &= a^* , \\ [a, X_0] &= -\text{Hess } V\, b , & [a^*, X_0] &= -b^*\, \text{Hess } V \end{aligned} \tag{5.33}$$

where we make use of the notations (5.15) and (5.16). We will often use this matricial notation where $*$ refers to forms or line matrices. As an example, we also have by combination the formulas:

$$[\Lambda^2, X_0] = -b^*(\text{Hess } V - \text{Id})a - a^*(\text{Hess } V - \text{Id})b \tag{5.34}$$

and

$$b(b^*b) = (b^*b + 1)b . \tag{5.35}$$

Remark 5.14.
Note that since we are working with pseudo-differential operators which all belong to classes associated with the metric $dx^2 + dv^2 + d\xi^2 + d\eta^2$, all the commutators are well defined as continuous operators from $\mathcal{S}$ to $\mathcal{S}$ or from $\mathcal{S}'$ to $\mathcal{S}'$.

5.3.4 Hypoelliptic Estimates: A Basic Lemma

As a consequence of these relations combined with the estimates[1]

$$\left\| \Lambda^{2\rho-2} a^* \right\| \leq 1 \tag{5.36}$$

and

$$\left\| \Lambda^{2\rho-2} b^* \right\| \leq 1 , \tag{5.37}$$

for $\rho \leq 1/2$, one has the following result which is adapted from Lemma 2.5 of [HerNi].

Lemma 5.15.
Take $\rho \in [0, 1/4]$. The estimate

$$\left\| \Lambda^{\rho} u \right\|^2 \leq \operatorname{Re} \langle K u \mid (L + L^*) u \rangle - \operatorname{Re} \langle \mathcal{A}^* b u \mid u \rangle + \operatorname{Re} \langle L K u \mid L u \rangle \\ - \operatorname{Re} \langle \mathcal{A}^* b u \mid L u \rangle + 3 \left\| b u \right\|^2 + 3 \left\| u \right\|^2 , \tag{5.38}$$

holds for any $u \in \mathcal{S}(\mathbb{R}^{2n})$, with

$$L = \Lambda^{2\rho-2} a^* b = \Lambda^{2\rho-2} \left(\sum_j a_j^* b_j \right)$$

and

$$\mathcal{A}^* = \left[\Lambda^{2\rho-2} a^*, X_0 \right] = \left(\mathcal{A}_j^* \right), \ \mathcal{A}_j^* = \left[\Lambda^{2\rho-2} a_j^*, X_0 \right] .$$

Let us give the proof for the sake of completeness.
Step 1.
We first show that:

$$\left\| \Lambda^{\rho} u \right\|^2 \leq \operatorname{Re} \langle X_0 u \mid (L + L^*) u \rangle - \operatorname{Re} \langle \mathcal{A}^* b u \mid u \rangle + \left\| b u \right\| \left\| u \right\| + \left\| \Lambda^{\rho-1} u \right\|^2 . \tag{5.39}$$

The starting point is

$$\left\| \Lambda^{\rho} u \right\|^2 = \langle \Lambda^{2\rho-2} b^* b u \mid u \rangle + \langle \Lambda^{2\rho-2} a^* a u \mid u \rangle + \langle \Lambda^{2\rho-2} u \mid u \rangle .$$

We obtain the result immediately from (5.37) and from the identity

$$\langle \Lambda^{2\rho-2} a^* a u \mid u \rangle = \operatorname{Re} \langle X_0 u \mid (L + L^*) u \rangle - \operatorname{Re} \langle \mathcal{A}^* b u \mid u \rangle ,$$

which simply results from $a = b X_0 - X_0 b$ (cf (5.33)).
Step 2.
We now show that

$$\operatorname{Re} \langle X_0 u \mid (L + L^*) u \rangle \leq \operatorname{Re} \langle K u \mid (L + L^*) u \rangle - 2 \operatorname{Re} \langle b^* b u \mid L u \rangle + \left\| b u \right\| \left\| u \right\| . \tag{5.40}$$

---

[1] Actually, we do not need for our qualitative presentation to have such a precise control.

We start from

$$\mathrm{Re}\,\langle X_0 u\mid (L+L^*)u\rangle = \mathrm{Re}\,\langle Ku\mid (L+L^*)u\rangle - \mathrm{Re}\,\langle b^*bu\mid Lu\rangle - \mathrm{Re}\,\langle b^*bu\mid L^*u\rangle\,,$$

and work on the last term of the right hand side. We have

$$\mathrm{Re}\,\langle b^*bu\mid L^*u\rangle = \mathrm{Re}\,\langle bb^*bu\mid a\Lambda^{2\rho-2}u\rangle\,.$$

Using $bb^*b = (b^*b + 1)\,b$ we get

$$\mathrm{Re}\,\langle b^*bu\mid L^*u\rangle = \mathrm{Re}\,\langle b^*bbu\mid a\Lambda^{2\rho-2}u\rangle + \mathrm{Re}\,\langle bu\mid a\Lambda^{2\rho-2}u\rangle\,.$$

The next point is to observe the commutation of b^*b with Λ

$$\begin{aligned}
\mathrm{Re}\,\langle b^*bu\mid L^*u\rangle &= \mathrm{Re}\,\langle bu,\mid a\Lambda^{2\rho-2}b^*bu\rangle + \mathrm{'Re}\,\langle bu\mid a\Lambda^{2\rho-2}u\rangle\\
&= \mathrm{Re}\,\langle b^*bu\mid Lu\rangle + \mathrm{Re}\,\langle bu\mid a\Lambda^{2\rho-2}u\rangle\,.
\end{aligned}$$

We conclude by using for the last term (5.36).

Step 3.

It remains to control $-2\,\mathrm{Re}\,\langle b^*bu\mid Lu\rangle$. We will show

$$-2\,\mathrm{Re}\,\langle b^*bu\mid Lu\rangle \le \frac{3}{2}||bu||^2 + \mathrm{Re}\,\langle LKu\mid Lu\rangle - \mathrm{Re}\,\langle \mathcal{A}^*bu\mid Lu\rangle + \frac{1}{2}||u||^2\,. \tag{5.41}$$

We start from

$$\begin{aligned}
-2\,\mathrm{Re}\,\langle b^*bu\mid Lu\rangle &\le ||bu||^2 + ||bLu||^2\\
&\le ||bu||^2 + \mathrm{Re}\,\langle KLu\mid Lu\rangle\\
&\le ||bu||^2 + \mathrm{Re}\,\langle [K,L]u\mid Lu\rangle + \mathrm{Re}\,\langle LKu\mid Lu\rangle\,.
\end{aligned}$$

We now observe that:

$$[K,L] = -L - \mathcal{A}^*b - \Lambda^{2\rho-2}a^*a\,.$$

The last term to control is

$$-\,\mathrm{Re}\,\langle \Lambda^{2\rho-2}a^*au\mid Lu\rangle = -\,\mathrm{Re}\,\langle a\Lambda^{4\rho-4}a^*au\mid bu\rangle\,.$$

Using again (5.36), it is controlled when $\rho \le \frac{1}{4}$.

Putting together (5.39), (5.40) and (5.41) ends the proof of the lemma.

5.3.5 Proof of Theorem 5.8

We are now able to prove Theorem 5.8 by bounding each of the six terms in the right-hand side of (5.38):

First term:

We write

$$|\langle Ku \mid Lu \rangle| \le \|Ku\| \, \|Lu\| \le \|Ku\| \, \left\|\Lambda^{2\rho-2}a^*\right\| \, \|bu\|$$

and recall $\left\|\Lambda^{2\rho-2}a^*\right\| \le 1$ for $\rho \le 1/2$. The simple inequality

$$\|bu\|^2 = \langle b^*bu \mid u \rangle = \mathrm{Re}\,\langle Ku \mid u \rangle \le \|Ku\| \, \|u\|$$

now gives

$$|\langle Ku \mid Lu \rangle| \le \|Ku\|^{3/2} \, \|u\|^{1/2} \,.$$

For the second part we write $|\langle Ku \mid L^*u \rangle| \le \|Ku\| \, \|L^*u\|$ and we use, observing the commutation of b^*b with a and Λ,

$$L^* = b^*a\Lambda^{2\rho-2}(1+b^*b)^{-1/2}(1+b^*b)^{1/2} = b^*(1+b^*b)^{-1/2}a\Lambda^{2\rho-2}(1+b^*b)^{1/2} \,.$$

From this we deduce

$$\|L^*u\| \le \left\|b^*(1+b^*b)^{-1/2}\right\| \, \left\|a\Lambda^{2\rho-2}\right\| \, \left\|(1+b^*b)^{1/2}u\right\|$$

$$\le C_n \left(\|Ku\| \, \|u\| + \|u\|^2\right)^{1/2} \,,$$

and we obtain

$$|\langle Ku \mid (L+L^*)u \rangle| \le C \left(\|Ku\|^2 + \|u\|^2\right) \,.$$

Terms 5 and 6:
They are all bounded by

$$C \left(\|bu\|^2 + \|u\|^2\right) \le C' \left(\|Ku\|^2 + \|u\|^2\right) \,.$$

Term 3:
We write

$$\mathrm{Re}\,\langle LKu \mid Lu \rangle = \mathrm{Re}\,\langle \Lambda^{2\rho-2}a^*bKu \mid \Lambda^{2\rho-2}a^*bu \rangle = \mathrm{Re}\,\langle a\Lambda^{4\rho-4}a^*bKu \mid bu \rangle \,.$$

Since a, a^* and b belong to $\mathrm{Op}\,S^1_{\Psi}$, the operator $a\Lambda^{4\rho-4}a^*b$ is bounded for $\rho \le 1/4$, which is just the condition appearing in Theorem 5.8. We get

$$|\,\mathrm{Re}\,\langle LKu \mid Lu \rangle| \le C \, \|Ku\| \, \|bu\| \le C \left(\|Ku\|^2 + \|u\|^2\right) \,,$$

for $\rho \le 1/4$.
Term 4:
We write

$$\mathrm{Re}\,\langle A^*bu \mid Lu \rangle = \mathrm{Re}\,\langle \left[\Lambda^{2\rho-2}a^*, X_0\right] bu \mid \Lambda^{2\rho-2}a^*bu \rangle$$

$$= \mathrm{Re}\,\langle a\Lambda^{2\rho-2}\left[\Lambda^{2\rho-2}a^*, X_0\right] bu \mid bu \rangle \,.$$

The hamiltonian vector field X_0 belongs to $\mathrm{Op}\,S^2_{\Psi}$ (see (5.14)) and the pseudo-differential calculus for commutators gives:

$$a\Lambda^{2\rho-2}\left[\Lambda^{2\rho-2}a^*, X_0\right] \in \operatorname{Op} S_\Psi^{1+4\rho-4+1+2-1} = \operatorname{Op} S_\Psi^{4\rho-1} \subset \mathcal{L}(L^2),$$

for $\rho \leq 1/4$. We conclude like for the third term with

$$|\operatorname{Re}\langle \mathcal{A}^* bu \mid Lu\rangle| \leq C\left(\|Ku\|^2 + \|u\|^2\right).$$

Term 2:
This term is the more delicate and we have to split the variables x and v while refining our pseudo-differential calculus with some exact commutator expressions. First we have

$$\begin{aligned}
\mathcal{A}^* &= \left[\Lambda^{2\rho-2}a^*, X_0\right] = \left[\Lambda^{2\rho-2}, X_0\right]a^* + \Lambda^{2\rho-2}b^*\operatorname{Hess} V \\
&= (b^*b+1)^{1/2}\left[\Lambda^{2\rho-2}a^*, (b^*b+1)^{-1/2}X_0\right]a^* \\
&\quad + (b^*b+1)^{1/2}\Lambda^{2\rho-2}(b^*b+1)^{-1/2}b^*\operatorname{Hess} V \\
&= (b^*b+1)^{1/2}(A_1 + A_2),
\end{aligned}$$

with

$$A_1 := (1+b^*b)^{-1/2}\left[\Lambda^{2\rho-2}, X_0\right],$$

and

$$A_2 := \Lambda^{2\rho-2}(b^*b+1)^{-1/2}b^*\operatorname{Hess} V.$$

If A_1 and A_2 are bounded, one obtains

$$|\operatorname{Re}\langle \mathcal{A}^* bu \mid u\rangle| \leq \left\|\operatorname{Re}\langle(A_1 + A_2)bu \mid (1+b^*b)^{1/2}u\rangle\right\| \leq C\left(\|Ku\|^2 + \|u\|^2\right).$$

The boundedness of A_2 is simple to verify. The coefficients $\Lambda^{2\rho-2}\partial^2_{x_i x_j}V$ belong to $\operatorname{Op} S_\Psi^{2\rho-2+1}$ and are bounded if $\rho \leq 1/2$. The boundedness of A_2 is then a consequence of the property that $(1+b^*b)^{-1/2}b^* \in \mathcal{L}(L^2)$.

Noting that $A_1 = \left((1+b^*b)^{-1/2}\left[\Lambda^{2\rho-2}, X_0\right]\Lambda\right)(\Lambda^{-1}a^*)$, the boundedness of A_1 is given by the following lemma applied with $r_1 = 0$, $r_2 = 2\rho - 2$ and $r_3 = 1$ (this requires $\rho \leq 1/2$).

Lemma 5.16.
For $r_1 + r_2 + r_3 \leq 0$, the operator $(1+b^*b)^{-1/2}\Lambda^{r_1}\left[\Lambda^{r_2}, X_0\right]\Lambda^{r_3}$ is bounded on $L^2(\mathbb{R}^{2n})$.

Proof:
Since $\Lambda^{r_1}\left[\Lambda^{r_2}, X_0\right] = \left[\Lambda^{r_1+r_2}, X_0\right] - \left[\Lambda^{r_1}, X_0\right]\Lambda^{r_2}$ we can simply consider the case $r_1 = 0$. Note that the vector field $X_0 = v \cdot \partial_x - \partial_x V(x) \cdot \partial_v$ is the sum of terms in the form $\ell(v, D_v)a(x, v, D_x, D_v)$ where ℓ is a linear symbol in (v, η) and $a \in S_\Psi^1$. We expand the commutator as

$$\begin{aligned}
\left[\Lambda^{r_2}, \ell(v, D_v)a(x, v, D_x, D_v)\right] \\
= \left[\Lambda^{r_2}, \ell(v, D_v)\right]a(x, v, D_x, D_v) + \ell(v, D_v)\left[\Lambda^{r_2}, a(x, v, D_x, D_v)\right] \\
:= B_1 + B_2.
\end{aligned}$$

Since the commutator $[\Lambda^{r_2}, a(x, v, D_x, D_v)]$ belongs to $\operatorname{Op} S_\Psi^{r_2}$, we have

$$(1 + b^*b)^{-1} B_2 \Lambda^{-r_2} \in \mathcal{L}(L^2) .$$

Let us now look at B_1. It is enough to show the

Sublemma 5.17.
$$[\Lambda^{r_2}, \ell(v, D_v)] \in \operatorname{Op} S_\Psi^{r_2-1} .$$

We first note that this is not a direct consequence of the previous pseudo-differential calculus which says only that this term is in $\operatorname{Op} S_\Psi^{r_2}$. But this calculus says that modulo $S_\Psi^{r_2-1}$ the symbol of the commutator is obtained by $\frac{1}{i}$ the Poisson bracket of the principal symbols of Λ^{r_2} (computed in Theorem 4.8) and of $\ell(v, D_v)$. An explicit computation based on (4.18) gives that this Poisson bracket is actually in $S_\Psi^{r_2-1}$. ∎

We proved
$$| \operatorname{Re} \langle \mathcal{A}^* bu \mid u \rangle | \le C \left(\| Ku \|^2 + \| u \|^2 \right) ,$$

which ends the proof of (5.21) in Theorem 5.8.

Another version of the estimate (5.21) can be written by following the same lines as in the proof of Theorem 5.8 if one notices that X_0 occurs only through commutators and that adding a term $i\nu$ to X_0 or K with $\nu \in \mathbb{R}$ does not change the real part of K. So Theorem 5.8 admits the following extension.

Theorem 5.18.
Under assumption (5.6) or (5.7) for V, there exists a constant $C > 0$ such that

$$\forall \nu \in \mathbb{R}, \ \forall u \in D(K), \ \left\| \Lambda^{1/4} u \right\|^2 \le C \left(\| (K - i\nu)u \|^2 + \| u \|^2 \right). \tag{5.42}$$

5.4 Necessary Conditions with Respect to the Corresponding Witten Laplacian

We recall that for any $V \in C^\infty(\mathbb{R}^n)$, the Laplacian $\Delta_{V/2}^{(0)}$ is essentially self-adjoint on $C_0^\infty(\mathbb{R}^n)$ or on $S(\mathbb{R}^n)$ if V is tempered (i.e. with all its derivatives polynomially bounded).
The operators $K = v \cdot \partial_x - \partial_x V(x) \cdot \partial_v - (\partial_v - v/2) \cdot (\partial_v + v/2) = X_0 + b^*b$ and $K_- = -X_0 + b^*b$ with domain $C_0^\infty(\mathbb{R}^{2n})$ (or $S(\mathbb{R}^{2n})$ if V is tempered) are accretive and closable, with maximally accretive closure. We use the same notation K for the closure of K.
We recall that $\lambda \in \mathbb{C}$ belongs to the essential spectrum of K if there exists a normalized sequence $(U_k)_{k\in\mathbb{N}}$ in $L^2(\mathbb{R}^{2n})$, with $U_k \in D(K)$, such that $\lim_{k\to\infty} \| (K - \lambda)U_k \| = 0$.

Proposition 5.19.
Assume that V is C^∞ function.

i) If K has a compact resolvent then the Witten Laplacian $\Delta_{V/2}^{(0)}$ has a compact resolvent.

ii) If 0 belongs to the essential spectrum of $\Delta_{V/2}^{(0)}$ then 0 is in the essential spectrum of K (and K_-).

Let us first consider i).
By contradiction, assume that $\Delta_{V/2}^{(0)}$ does not have a compact resolvent. Then there exists an orthonormal sequence $(u_k)_{k \in \mathbb{N}}$ such that

$$\langle u_k \mid \Delta_{V/2}^{(0)} u_k \rangle = \|au_k\|^2$$

is bounded. The sequence given by

$$U_k(x, v) = u_k(x)(2\pi)^{-n/4}e^{-v^2/4}$$

is orthonormal and satisfies

$$\forall k \in \mathbb{N}, \ KU_k = au_k \otimes (2\pi)^{-n/4}ve^{-v^2/4} \quad \text{in } \mathcal{D}'(\mathbb{R}^{2n}) \ .$$

Since $K = K_-^*$, each function U_k belongs to $D(K)$ and the sequence KU_k is bounded. If K has a compact resolvent, we could extract a Cauchy subsequence U_k. This immediately implies that u_k should be a Cauchy sequence in $L^2(\mathbb{R}^n)$. But this is in contradiction with the fact that u_k is an orthonormal sequence.
For ii), we assume $\lim_{k \to \infty} \langle u_k \mid \Delta_{V/2}^{(0)} u_k \rangle = \lim_{k \to \infty} \|au_k\|^2 = 0$ and the consequence $\lim_{k \to \infty} \|KU_k\| = 0$ says that 0 belongs to the essential spectrum of K.

Remark 5.20.
The results of Proposition 5.19 tell us that it is not possible to derive directly the compactness of the resolvent of the Fokker-Planck operator $K = X_0 + b^*b$ from the one of $-\Delta_x + \frac{1}{4}|\nabla V(x)|^2$. We will indeed prove (see Subsection 11.3.1) that $\Delta_{V/2}^{(0)}$ has not a compact resolvent for $V(x_1, x_2) = x_1^2 x_2^2$ contrary to $-\Delta_x + \frac{1}{4}|\nabla V(x)|^2$. As a consequence a naive application of Kohn's method as presented in Section 2.3 would surely fail and one has to introduce in the analysis the operators a and a^* or by some other mean the specific structure of the Witten Laplacian $\Delta_{V/2}^{(0)}$.

5.5 Analysis of the Fokker-Planck Quadratic Model

In the case when the potential V is quadratic, we will see in Subsection 5.5.1 that the spectrum can be explicitly computed. One should not overestimate the interest of explicit computations. The operator being diagonalized (in the generic case) in a non orthonormal basis, this does not lead to good estimates for the resolvent which have to be proven in a different way. Nethertheless, we shall show in Subsection 5.5.2 how one can improve in the quadratic case the estimates obtained in Theorem 5.8.

5.5.1 Explicit Computation of the Spectrum

We follow here Risken [Ris], who refers actually to [Brin]. We consider the case when $n = 1$ and $V(x) = \omega_0^2 \frac{x^2}{2}$. After a dilation in the x variable we have consequently to analyze the operator:

$$L = -\frac{d^2}{dv^2} + \frac{1}{4}v^2 - \frac{1}{2} - \omega_0(v\partial_x - x\partial_v) ,\qquad(5.43)$$

with $\omega_0 \neq 0$.
With

$$b = \partial_v + \frac{1}{2}v , \ a = \partial_x + \frac{1}{2}x ,$$

this can also be written as:

$$L = b^*b + \omega_0(b^*a - a^*b) .\qquad(5.44)$$

The trick is to see the operator as a "complex" harmonic oscillator .
We are looking for an expression of the type:

$$L = \lambda_1 c_{1,+}c_{1,-} + \lambda_2 c_{2,+}c_{2,-}\qquad(5.45)$$

where λ_1 and λ_2 are complex numbers and $c_{1,-}$, $c_{1,+}$, $c_{2,-}$ and $c_{2,+}$ satisfy standard commutation relations:

$$\begin{aligned}[c_{1,-} , c_{1,+}] = [c_{2,-} , c_{2,+}] = 1 , \\ [c_{1,-} , c_{2,+}] = [c_{1,+} , c_{2,-}] = [c_{1,-} , c_{2,-}] = [c_{1,+} , c_{2,+}] = 0 ,\end{aligned}\qquad(5.46)$$

and the other equation:

$$[L , c_{i,\pm}] = \mp c_{i,\pm} .\qquad(5.47)$$

More explicitly,

$$\begin{aligned}c_{1,+} &= \delta^{-\frac{1}{2}}\left(\sqrt{\lambda_1}b^* - \sqrt{\lambda_2}a^*\right) , \\ c_{1,-} &= \delta^{-\frac{1}{2}}\left(\sqrt{\lambda_1}b + \sqrt{\lambda_2}a\right) , \\ c_{2,+} &= \delta^{-\frac{1}{2}}\left(-\sqrt{\lambda_2}b^* + \sqrt{\lambda_1}a^*\right) , \\ c_{2,-} &= \delta^{-\frac{1}{2}}\left(\sqrt{\lambda_2}b + \sqrt{\lambda_1}a\right) .\end{aligned}\qquad(5.48)$$

Here

$$\delta = \sqrt{1 - 4\omega_0^2} ,$$

(which is assumed to be different from 0) and

$$\lambda_1 = (1 + \delta)/2 \,, \ \lambda_2 = (1 - \delta)/2 \,.$$

We emphasize that in general $c_{j,-}$ IS NOT the formal adjoint of $c_{j,+}$ but nethertheless the construction of the eigenvectors, by use of "creation operator", is working. One obtains actually a complete system of eigenvectors by introducing:

$$\psi_{n_1,n_2} = (n_1!\,n_2!)^{-\frac{1}{2}}(c_{1,+})^{n_1}(c_{2,+})^{n_2}\psi_{0,0}\,,$$

where

$$\psi_{0,0} = (1/2\pi)^{\frac{1}{2}} \exp -\frac{1}{4}(x^2 + v^2)\,.$$

The corresponding eigenvalue is:

$$\lambda_{n_1,n_2} = \lambda_1 n_1 + \lambda_2 n_2\,. \tag{5.49}$$

So there are mainly two cases according to the sign of $(1 - 4\omega_0^2)$ and a special case corresponding to $\omega_0 = \pm\frac{1}{2}$. We emphasize that these various cases do not appear in the discussion of the compactness of the resolvent.

When $0 < |\omega_0| < \frac{1}{2}$, the eigenvalues λ_j are real, so the spectrum of the Fokker-Planck operator is real.

When $|\omega_0| > \frac{1}{2}$, the spectrum is contained in a strictly convex sector in $\mathbb{C}$.

In the special case, $\lambda_1 = \lambda_2$. The previous method does not work !!

Remark 5.21.

Another approach is proposed in Risken [Ris]. It consists in expanding a function u in $L^2(\mathbb{R}^2)$ in the basis of the eigenfunctions of the harmonic oscillator, that is in the basis of the usual Hermite functions $h_{k_1,k_2}(x_1, x_2) = h_{k_1}(x_1)h_{k_2}(x_2)$ in two variables, and to observe that, for a given N, the spaces V_N generated by the h_{k_1,k_2} with $k_1 + k_2 = N$ are stable. We then have to analyze the restriction of the operator L to each V_N, that is a $(N+1) \times (N+1)$ matrix, whose eigenvalues can be explicitly computed.

The eigenvalue equation takes the form, for $y = (y_0, \ldots, y_N)$ and with the convention that $y_{-1} = y_{N+1} = 0$,

$$n\omega_0 y_{n-1} + (n - \lambda)y_n - (N - n)\omega_0 y_{n+1} = 0\,, \tag{5.50}$$

for $n = 0, \ldots, N$.

In the generic case, the eigenvalues are given by (5.49), with the additional condition $n_1 + n_2 = N$:

$$\lambda_{n_1,n_2} = \frac{N}{2} + \frac{1}{2}\delta(n_1 - n_2)\,. \tag{5.51}$$

In the case when $\omega_0 = \pm\frac{1}{2}$, the matrix has a unique eigenvalue $\frac{N}{2}$ (with algebraic multiplicity $(N + 1)$). One can only write a Jordan form.

Note that this example has some connection with a model also discussed by Davies [Dav5] and later in [Sj6].

5.5.2 Improved Estimates for the Quadratic Potential

We will show by some more explicit method that the lower bound (5.21) can be improved when the potential is quadratic.

Proposition 5.22.
Let V be a polynomial real potential of degree less or equal to 2 with

$$\det(\operatorname{Hess} V) \neq 0 .$$

Let K denote the maximally accretive operator

$$K := v \cdot \partial_x - \partial_x V(x) \cdot \partial_v - \Delta_v + \frac{v^2}{4} - n/2 ,$$

and set

$$\Lambda_x = (-\Delta_x^2 + x^2/4)^{1/2} , \ \Lambda_v = (-\Delta_v^2 + v^2/4)^{1/2} .$$

Then there exists a constant $C > 0$ such that

$$\forall u \in D(K), \ \left\| \Lambda_x^{2/3} u \right\|^2 + \|\Lambda_v u\|^2 \leq C \left(\|Ku\|^2 + \|u\|^2 \right) .$$

We shall first consider for $X = (x, \xi) \in \mathbb{R}^{2n}$ the operator

$$K_X = i\xi \cdot v - x \cdot \partial_v - \Delta_v + \frac{v^2}{4} - n/2$$

acting on $L^2(\mathbb{R}_v^n)$. After conjugation with a unitary dilation it equals

$$K'_{X'} = i\xi' \cdot v - x' \cdot \partial_v - \frac{1}{2} \left(-\Delta_v + v^2 - n \right) ,$$

with $X' = (x', \xi') = (2^{-1/2}x, 2^{1/2}\xi)$. Let H_k denote the harmonic oscillator $\frac{1}{2} \left(-\Delta_{v_k} + v_k^2 \right)$. Then, for any $a \in \mathcal{S}'(\mathbb{R}^{2n})$, we have

$$e^{it_k H_k} a(v, D_v) e^{-it_k H_k} =$$
$$a(\ldots, v_k \cos(t_k) - D_{v_k} \sin(t_k), v_k \sin(t_k) + D_{v_k} \cos(t_k), \ldots) ,$$

where the dots stand for unchanged variables. We use it with $a(v, D_v) = K'_{X'}$ and we obtain

$$\Pi_{k=1}^n e^{-it_k H_k} K'_{X'} \Pi_{k=1}^n e^{it_k H_k} = K'_{X''(t)}$$

with $t = (t_1, \ldots, t_n)$, $X''(t) = (x''(t), \xi''(t))$ and

$$\begin{cases} x''(t)_k = x'_k \cos(t_k) - \xi'_k \sin(t_k) \\ \xi''(t)_k = x'_k \sin(t_k) + \xi'_k \cos(t_k) \end{cases}$$

For any $k \in \{1, \ldots, n\}$, we choose t_k so that

$$x''(t_k)_k = 0 \quad \text{and} \quad \xi''(t_k)_k = \sqrt{|x'_k|^2 + |\xi'_k|^2} = \rho_k .$$

With this choice we obtain

$$K''_{X''(t)} = \frac{1}{2} \left(\sum_{k=1}^n \left(i\rho_k v_k - \partial_{v_k}^2 + v_k^2 \right) - n \right) .$$

After a metaplectic transformation associated to a rotation in the v-variable, we arrive to

$$\hat{K}^{(3)} = \frac{1}{2} \left(\rho v_1 - \Delta_v + v^2 - n \right) . \tag{5.52}$$

$$\rho = \sqrt{\sum_k \rho_k^2} .$$

After another unitary transform we obtain

$$K^{(4)} = \frac{1}{h} \left(\frac{i}{2} v_1 - h^2 \Delta_v + \frac{v^2}{4} - \frac{nh}{2} \right) ,$$

with

$$h = \rho^{-2} \quad \text{and} \quad \rho = \sqrt{\sum_k \rho_k^2}.$$

The problem is then reduced to the analysis of the operator

$$-h^2 \Delta + \frac{1}{4} v^2 + \frac{i}{2} v_1 , \quad v \in \mathbb{R}^n$$

for which a semi-classical version (cf [DeSjZw], in particular Figure 1.1) of a classical theorem on principal type differential subelliptic operators (See Chapter 27 in [Hor2]) of the form $L_1 + iL_2$, with $[L_2, [L_2, L_1]] \neq 0$, can be applied.
The symbol $(\eta^2 + \frac{1}{4} v^2) + i\frac{1}{2} v_1$ vanishes only at the point $(v, \eta) = (0, 0)$, and we have at this point:

$$\{v_1, (\frac{1}{4} v^2 + \eta^2)\} = 0 \quad \text{and} \quad \{v_1, \{v_1, \eta^2 + \frac{1}{4} v^2\}\} = 2 \neq 0.$$

The result of [DeSjZw] gives the estimate[2]:

$$\forall u \in \mathcal{C}_0^\infty(\mathbb{R}^n), \quad h^{2/3} \|u\| \leq C \left\| (-h^2 \Delta_v + v^2 + iv_1) u \right\| .$$

After the unitary transforms, we obtain for some constant $C_1 > 0$

$$\forall u \in \mathcal{C}_0^\infty(\mathbb{R}^n_v), |\rho|^{4/3} \|u\|^2 \leq C_1 \left(\left\| K''_{X''(t)} u \right\|^2 + \|u\|^2 \right) ,$$

---

[2] Another approach will be introduced later in Chapter 9.

hence

$$\forall u \in D(K_X), (1 + |X|^2)^{2/3} \|u\|^2 \leq C_1 \left(\|K_X u\|^2 + \|u\|^2 \right) . \qquad (5.53)$$

We now consider the operator $K = v.\partial_x - \partial_x V(x).\partial_v - \Delta_v^2 + v^2/4 - n/2$ where V is a polynomial of degree not greater than 2. After diagonalizing the quadratic part and possibly using a unitary transform U defined by

$$f \mapsto Uf(x,v) = e^{i\eta_0 v} f(x - x_0) ,$$

it suffices to consider the case

$$K = \sum_{k=1}^{n} \left(v_k \partial_{x_k} - \lambda_k x_k \partial_{v_k} - \partial_{v_k}^2 + \frac{1}{4} v_k^2 \right) - C_0 .$$

The operator $(1 + K^*K)$ has the following form

$$(1 + K^*K) = \sum_j Q_j(x, D_x) P_j(v, D_v) + (-\Delta^2 + v^2/4)^2 - C_0' ,$$

where the Q_j's and P_j's are polynomials of degree less or equal to 2.

We now use the anti-Wick quantization and its comparison with the Weyl quantization . This comparison is given by (4.24) in the calculus associated with the metric $(1 + |x|^2 + |\xi|^2)^{-1}(dx^2 + d\xi^2)$ in which quadratic symbols enter naturally (We apply here the relation (4.24) in dimension $d = n$ with $z = x$ and $\zeta = \xi$). It leads to

$$(1 + K^*K) \geq \sum_j Q_j^{A-Wick}(x, D_x) P_j(v, D_v)$$

$$+ (-\Delta^2 + v^2 - n/2)^2 - C_3(-\Delta_v + v^2/4)$$

$$\geq (1 + K_{X_\lambda}^* K_{X_\lambda})^{A-Wick,x} - C_4(-\Delta_v + v^2/4) ,$$

where $X_\lambda = (\lambda_1 x_1, \ldots, \lambda_n x_n, \xi)$, where the constants C_3 and C_4 depend on the second derivatives of $V(x)$ and where the superscript " $A - Wick, x$ " refers to the anti-Wick quantization with respect to the x variable only. By setting $\lambda_{4/3}(x, \xi) = \left(|\xi|^2 + x^2/4 + 1 \right)^{2/3}$, the comparison of Weyl and anti-Wick quantization (4.24) yields

$$(-\Delta_x + x^2/4)^{2/3} = \lambda_{4/3}^{A-Wick}(x, D_x) + R, \quad \text{with } R \in \mathcal{L}(L^2(\mathbb{R}^n)) .$$

Now since the anti-Wick quantization is a positive quantization (it associates to a positive symbol a positive operator) the estimate (5.53) gives

$$C_5(1 + K^*K) \geq (-\Delta_x + x^2/4)^{2/3} - C_5 \left(-\Delta_v + v^2/4 \right) .$$

We conclude with the inequality:

$$\|u\|^2 + \|Ku\|^2 \geq \text{Re} \langle Ku \mid u \rangle \geq \langle \left(-\Delta_v + \frac{1}{4} v^2 - \frac{n}{2} \right) u \mid u \rangle, \quad \forall u \in C_0^\infty(\mathbb{R}^{2n}) .$$

■

6

Return to Equilibrium
for the Fokker-Planck Operator

6.1 Abstract Analysis

In the case when A is a self-adjoint maximally accretive operator, the compactness of the resolvent suffices to give exponential return to equilibrium, which means here that by denoting Π_0 the spectral projection[1] on $\operatorname{Ker} A$,

$$\exists C, \alpha_1 > 0, \forall t \geq 0 \quad \left\| e^{-tA} - \Pi_0 \right\| \leq C e^{-\alpha_1 t} .$$

This is the case of the Witten Laplacian which is (after conjugation by a unitary operator) the generator of the Feller semi-group associated with the corresponding Dirichlet form. Another situation which is standard is when ellipticity leads to a sectorial operator with sector included in the set $\{z \in \mathbb{C}, \arg(z) \in [-\theta_0, \theta_0]\}$, with $\theta_0 < \pi/2$. In the case of the Fokker-Planck operator K, we have a non self-adjoint operator which is moreover not elliptic. Hence its numerical range, $\{\langle u, Ku \rangle, u \in D(K)\}$, is a priori the full closed half-plane $\{z \in \mathbb{C}, \operatorname{Re} z \geq 0\}$. However the hypoelliptic estimate can help for proving exponential return to equilibrium.

We summarize here in a pure functional analysis framework some of the arguments used in [HerNi] and [EckHai2], but we will not cover all the quantitative analysis which is given in [HerNi].

In a separable Hilbert space $\mathcal{H}$ we consider a maximally accretive operator $(K, D(K))$ such that

$$\sigma(K) \cap i\mathbb{R} = \sigma_{disc}(K) \cap i\mathbb{R} \subset \{0\} . \tag{6.1}$$

The spectral projection on its kernel will be denoted Π_0. We assume that there exist a self-adjoint operator $(\Lambda, D(\Lambda))$, $\Lambda \geq 1$, and three constants $C > 0$ and $M \geq m > 0$ such that

$$\|\Lambda^m u\|^2 \leq C \left(\|(K - i\nu)u\|^2 + \|u\|^2 \right) , \forall \nu \in \mathbb{R}, \forall u \in D(K) , \tag{6.2}$$

---

[1] When $\operatorname{Ker} A = 0$, it should be understood that $\Pi_0 = 0$.

$$\|Ku\|^2 \le C \left\|A^M u\right\|^2 , \forall u \in D(A^M) . \tag{6.3}$$

In the applications A is associated with some Sobolev scale. The second estimate says that K is a finite order operator in this scale, (in the case of the Fokker-Planck operator, $K \in \operatorname{Op} S_\Psi^2$ and $M = 2$). The first one is an hypoelliptic estimate as stated in Theorem 5.18 with $m = 1/4$.

Theorem 6.1.
If K is a maximally accretive operator on $\mathcal{H}$ which satisfies (6.1),(6.2) and (6.3), then there exist two positive constants $C, \alpha_1 > 0$ such that

$$\forall t \ge 0, \quad \left\|e^{-tK} - \Pi_0\right\| \le Ce^{-\alpha_1 t}.$$

As a consequence of Von Neumann theorem (see [RiNa]), for any maximally accretive operator K, the semi-group e^{-tK} can be written as the weakly (consider $\langle e^{-tK} u_0 \mid \varphi \rangle$ with $u_0 \in \mathcal{H}$ and $\varphi \in D(K^*)$) convergent integral

$$e^{-tK} = \frac{1}{2i\pi} \int_{+i\infty}^{-i\infty} e^{-tz}(z - K)^{-1} \, dz .$$

In order to prove the exponential return to equilibrium it suffices to deform the contour $[+i\infty, -i\infty]$ into a contour essentially inside $\{\operatorname{Re} z > 0\}$ with some algebraic positive lower bound of $\operatorname{Re} z$ and some algebraic upper bound of $\left\|(z - K)^{-1}\right\|$ with respect to $|z|$. Then we can write according to the next picture

$$e^{-tK} = \frac{1}{2i\pi} \int_{\partial S_K} e^{-tz}(z - K)^{-1} \, dz$$

and

$$e^{-tK} - \Pi_0 = \frac{1}{2i\pi} \int_{\partial S_K'} e^{-tz}(z - K)^{-1} \, dz .$$

Hence we are led to localize the spectrum of K and control the norm of its resolvent within its numerical range. This enters in the more general problem of localizing the pseudospectrum. We recall that pseudospectral estimates require the introduction a small parameter:

Definition 6.2.
For a closed operator $(A, D(A))$ on a Hilbert space $\mathcal{H}$, and for $\varepsilon > 0$, the ε-spectrum is defined by

$$\varepsilon\sigma(A) = \left\{ z \in \mathbb{C}, \ \left\|(z - A)^{-1}\right\| \le \frac{1}{\varepsilon} \right\} .$$

For any $\varepsilon > 0$, the ε-spectrum contains $\sigma(A)$ and is stable with respect to perturbations even when A is not self-adjoint nor normal. The information that it contains can be quite accurate especially when one considers some

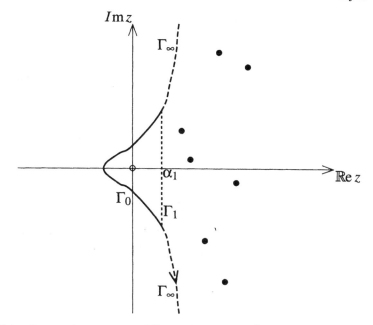

Fig. 6.1. Integration contour: $\partial S_K = \Gamma_0 \cup \Gamma_\infty$, $\partial S'_K = \Gamma_1 \cup \Gamma_\infty$, oriented from $\mathrm{Im}\, z = +\infty$ to $\mathrm{Im}\, z = -\infty$.

specific families $(A_\varepsilon)_{\varepsilon>0}$ of operators which depend on ε. With some variants, the pseudospectrum is then defined as $\cap_{\varepsilon>0}\varepsilon\sigma(A_\varepsilon)$.

Motivated by problems in numerical analysis and the analysis of non self-adjoint spectral problems this has received recent interest in the semiclassical regime (see [Bou], [Dav3], [Dav4], [Tref], [Zwo] and [DeSjZw] for further details and references).

Here we are interested in the large $|z|$ regime and we set $z = L\zeta$ with $\zeta \in \mathbb{C}$ bounded and $L \to \infty$. We are looking for an estimate of the form

$$\forall \zeta \in \mathbb{C}, 1/2 < |\zeta| \leq 1, \left\| (\zeta - L^{-1}K)^{-1} \right\| \leq CL^{N_0} ,$$

with $C, N_0 > 0$ independent of (L, ζ).

Theorem 6.1 is proved in three steps:

First step:
It is a simple consequence of the hypoellipticity assumption (6.2). We write, for $z = \mu + i\nu \in \mathbb{C}$, $\mu = \mathrm{Re}\, z \geq -1/2$, and $u \in D(K)$,

$$
\begin{aligned}
\|\Lambda^m u\|^2 &\leq C \left(\|(K - i\nu)u\|^2 + \|u\|^2 \right) \\
&\leq C \left(2\,\|(K - z)u\|^2 + (2\mu^2 + 1)\,\|u\|^2 \right) .
\end{aligned}
$$

We get

$$\|\Lambda^m u\|^2 \leq 2C \left(\|(K-z)u\|^2 + (\operatorname{Re} z + 1)^2 \|u\|^2 \right),$$
$$\forall z \in \mathbb{C}, \ \operatorname{Re} z \geq -1/2, \ \forall u \in D(K). \tag{6.4}$$

Second step:
The following lemma holds for any maximally accretive operator. The proof can be found in [HerNi].

Lemma 6.3.
Let $(K, D(K))$ be a maximally accretive operator in the Hilbert space $\mathcal{H}$. For any $\eta \in]0, 1[$, the estimate

$$|z+1|^{2\eta} \|u\|^2 \leq 4 \langle \, ((K+1)^*(K+1))^\eta \, u \mid u \, \rangle + 4 \, \|(z-K)u\|^2$$

holds for all $u \in D(K)$ and $z \in \mathbb{C}$ with $\operatorname{Re} z \geq -1$.

Although the proof does not follow any usual interpolation scheme, it can be viewed as an interpolation of the inequalities valid with coefficients 2 instead of 4 for $\eta \in \{0, 1\}$.

Third step:
Assumption (6.3) leads to the inequality

$$0 \leq (1+K^*)(1+K) \leq (1+\sqrt{C})^2 \Lambda^{2M},$$

and, according to the monotonicity of the operator functional $A \to A^\alpha$ for $\alpha \in [0, 1]$, to

$$0 \leq ((1+K^*)(1+K))^{\frac{m}{M}} \leq (1+\sqrt{C})^{\frac{2m}{M}} \Lambda^{2m}.$$

We apply the Lemma 6.3 with $\eta = \frac{m}{M}$ and obtain, for $\operatorname{Re} z \geq -1/2$ and $u \in D(K)$,

$$|z+1|^{2m/M} \|u\|^2 \leq 4(1+\sqrt{C})^{\frac{2m}{M}} \|\Lambda^m u\|^2 + 4 \|(K-z)u\|^2.$$

With the inequality (6.4) we obtain

$$\forall z \in \mathbb{C}, \ \operatorname{Re} z \geq -1/2, \ \forall u \in D(K),$$
$$|z+1|^{\frac{2m}{M}} \|u\|^2 \leq C_1 \|(K-z)u\|^2 + C_2 (\operatorname{Re} z + 1)^2 \|u\|^2,$$

with $C_1 = (8C(1+\sqrt{C})^{\frac{2m}{M}} + 4)$ and $C_2 = 4C(1+\sqrt{C})^{\frac{2m}{M}}$.
By taking $C' = \sqrt{2C_2}$ and $C'' = \sqrt{C_2/C_1}$, we deduce that the spectrum $\sigma(K)$ satisfies

$$\sigma(K) \subset S_K \cap (\{\operatorname{Re} z > 0\} \cup \{0\}), \tag{6.5}$$

where S_K is a $\mathcal{C}^\infty$ domain included in

$$S_K = \left\{ z \in \mathbb{C}, \ \operatorname{Re} z \geq -1/2, \ |z+1|^{m/M} \leq C'(\operatorname{Re} z + 1) \right\}.$$

Moreover we have the resolvent estimate

$$\forall z \in \overline{\mathbb{C} \setminus S_K},\ \mathrm{Re}\, z \geq -1/2,\quad \left\| (z-K)^{-1} \right\| \leq C'' \left| z+1 \right|^{-m/M}.$$

The assumption (6.1) says that there is no spectrum on $i\mathbb{R}$ except the possible eigenvalue 0 which is isolated and with finite multiplicity. Therefore it is possible to find an $\alpha_1 > 0$ which makes possible the contour deformation from ∂S_K to $\partial S'_K$, where Π_0 arises as the remaining residue.

6.2 Applications to the Fokker-Planck Operator

We conclude this chapter by giving explicit translations of Theorem 6.1 for the Fokker-Planck operator and its adjoint under the weak ellipticity assumption of Subsection 5.3.2. Note that, when $\exp -V$ is in L^1, we have

$$\mathrm{Ker}\, K = \mathbb{C}e^{-1/2\left(v^2/2 + V(x)\right)},$$

while this kernel is 0 if V takes negative values near infinity. In any case, $\mathrm{Ker}\, K = \mathrm{Ker}\, K^*$ and the spectral projection Π_0 is orthogonal. We shall denote by $M(x,v)$ the Maxwellian $M(x,v) = e^{-(v^2/2 + V(x))}$ and we write when $M \in L^1(\mathbb{R}^{2n})$

$$\forall u_0 \in L^2(\mathbb{R}^{2n}),\quad \Pi_0 u_0 = \frac{\int_{\mathbb{R}^{2n}} M^{1/2}(x,v) u(x,v)\, dx\, dv}{\int_{\mathbb{R}^{2n}} M(x,v)\, dx\, dv}.$$

We first give a result which holds when K has a compact resolvent. An extension will be sketched in the forthcoming subsection.

Theorem 6.4.
Assume that the potential $V(x)$ satisfies Assumption 5.6 or 5.7. Assume additionally in the second case that the Witten Laplacian $\Delta^{(0)}_{V/2}$ has a compact resolvent. Let $K = X_0 + b^*b$ be the maximally accretive operator defined by (5.2). Then there exists $\alpha_1 > 0$ and $C > 0$ so that

$$\forall u_0 \in L^2(\mathbb{R}^{2n}),\quad \left\| e^{-tK} u_0 - \Pi_0 u_0 \right\|_{L^2} \leq Ce^{-\alpha_1 t} \left\| u_0 \right\|. \tag{6.6}$$

Proof.
According to Corollary 5.10, the Fokker-Planck operator K has a compact resolvent. We have only to check that $Ku = i\lambda u$ with $\lambda \in \mathbb{R}$ implies $\lambda = 0$. The relation $\langle Ku \mid u \rangle = 0$ implies $u = \varphi(x)e^{-v^2/4}$ and $Ku = \sum_{i=1}^{n} (a\varphi)\, v_i e^{-v^2/4}$ which is possible only if $a\varphi = 0$ and $\lambda = 0$.

Corollary 6.5.
Under the assumption of Theorem 6.4, the same result holds for

$$K^* = -X_0 + b^*b$$

with the same constants:

$$\forall u_0 \in L^2(\mathbb{R}^{2n}),\quad \left\| e^{-tK^*} u_0 - \Pi_0 u_0 \right\|_{L^2} \leq Ce^{-\alpha_1 t} \left\| u_0 \right\|. \tag{6.7}$$

It suffices to notice

$$K^* = UKU^* \, ,$$

where U is the unitary transform given by $Uu(x,v) = u(x,-v)$.   ∎

Remark 6.6.
It is actually possible (see [HerNi]):

1. to prove also that e^{-tK} sends $\mathcal{S}'(\mathbb{R}^{2n})$ into $\mathcal{S}(\mathbb{R}^{2n})$ and that the exponential decay holds for any initial data $u_0 \in \mathcal{S}'(\mathbb{R}^{2n})$ up to some algebraic factor in $(t, 1/t)$;
2. to derive from this exponential return to equilibrium, the exponential decay of some relative entropy associated with the probability mesure $\left(\int M\right)^{-1} M$ when $V(x)$ is positive near infinity;
3. to get sharp lower and upper bounds for the constant α_1 in terms of the physical constants usually introduced with this equation, particle mass, temperature and friction coefficient.

6.3 Return to Equilibrium Without Compact Resolvent

We already said in Remark 5.13 that Assumption 5.7 allows to consider cases like $V(x) = \langle x \rangle$ for which the hypoelliptic estimate (5.21) holds without the compactness of the resolvent for $\Delta_{V/2}^{(0)}$, $\Delta_{\Phi/2}^{(0)}$ and K. Nevertheless if 0 belongs to the discrete spectrum of $\Delta_{V/2}^{(0)}$, it is still possible to prove

$$\sigma(K) \cap i\mathbb{R} = \sigma_{disc}(K) \cap i\mathbb{R} \subset \{0\} \, ,$$

and therefore to get an exponential return to equilibrium. It cannot be done simply by functional analysis arguments since we do not know how to prove directly that $K + i\lambda$, for $\lambda \in \mathbb{R}$, has a closed range. It is actually possible to adapt the method of [HerNi]-Section 5, for proving that there is no spectrum except possibly 0 in $\{\operatorname{Re} z \leq \alpha_1\}$ for some $\alpha_1 > 0$. Let us sketch the arguments:

First step:
One uses an hypoelliptic estimate like (5.38) with Λ replaced by Λ_δ, $\Lambda_\delta^2 = \delta^2 + a^*a + b^*b$, and with an explicit control with respect to $\delta > 0$ of the right-hand side. These estimates rely on the algebraic properties associated with the Fokker-Planck equation, the existence of a good pseudo-differential calculus (especially Lemma 5.16 is required) and the estimate[2]

$$\sup_{x \in \mathbb{R}^n} \max \sigma \left[(\operatorname{Hess}(V))^2(x) - (\frac{1}{4}|\nabla V(x)|^2 - \frac{1}{2}\Delta V(x))\operatorname{Id} \right] < +\infty \, ,$$

---

[2] $\max \sigma(A)$ denotes the highest eigenvalue of the matrix A.

which is satisfied when the condition (5.19) of Assumption 5.7 is replaced by

$$\forall \alpha \in \mathbb{N}^n, \ |\alpha| = 2, \ \forall x \in \mathbb{R}^n, \ |\partial_x^\alpha V(x)| \le C_\alpha (1 + |\nabla V(x)|) \langle x \rangle^{-\rho_0} \ .$$

Second step:
One optimizes the parameter δ by making use of the orthogonal decomposition written with $E_0 = \operatorname{Ker} K = \operatorname{Ker} \left[\Delta_{V/2}^{(0)} \right] \otimes \mathbb{C} e^{-v^2/4}$,

$$K = K\Big|_{E_0} \overset{\perp}{\oplus} K\Big|_{E_0^\perp} \qquad K^* = K^*\Big|_{E_0} \overset{\perp}{\oplus} K^*\Big|_{E_0^\perp}$$

$$\text{and } \Lambda_\delta = \Lambda_\delta\Big|_{E_0} \overset{\perp}{\oplus} \Lambda_\delta\Big|_{E_0^\perp} \ , \quad \text{with } \Lambda_\delta^2 = \delta^2 + a^* a + b^* b \ .$$

As an example, one obtains with this method the exponential return to equilibrium for any polyhomogeneous potential for which the principal part is $|x| \varphi(\frac{x}{|x|})$ and φ does not vanish on S^{n-1} (see Section 11.3).[3]

6.4 On Other Links Between Fokker-Planck Operators and Witten Laplacians

As used in an essential way in [HerNi], there are further links between the two operators. This will be done more quantitatively in Chapter 17 at the end of this book but let us just give as illustrating example some relations between α_1 and the lowest non zero eigenvalue ω_1 of the Witten Laplacian (assuming it exists).

If g_0 is chosen as

$$g_0 = \Psi(x)(2\pi)^{-\frac{d}{4}} \exp -\frac{v^2}{4}$$

where $\Psi(x)$ is a normalized eigenfunction of $\Delta_{\frac{V}{2}}^{(0)}$ corresponding to the eigenvalue $\omega_1 > 0$, then we observe (using (6.6) and (6.7)) the following inequality

$$\begin{aligned}
\tfrac{d}{dt} \| \exp -tK g_0 \|^2 &= -2 \operatorname{Re} \langle K \exp -tK g_0 \mid \exp -tK g_0 \rangle \\
&\ge -2 \| K g_0 \| \, \| \exp -tK^* \exp -tK g_0 \| \\
&\ge -2\sqrt{d+2}\sqrt{\omega_1} C^2 \exp -2\alpha_1 t \ .
\end{aligned} \qquad (6.8)$$

Integrating between 0 and t, leads to

$$1 - \| \exp -tK g_0 \|^2 \le \frac{1}{\alpha_1} \sqrt{d+2}\sqrt{\omega_1} C^2 (1 - \exp -2\alpha_1 t) \ .$$

Taking the limit $t \to +\infty$, this leads to

---

[3] We thank C. Villani who asked the question about this case which is missing in [HerNi].

$$\alpha_1 \leq \sqrt{d+2} \, \sqrt{\omega_1} \, C^2 \ .$$

Of course it is only interesting when controlling the constant C as it is done in [HerNi]. One expects that in the semi-classical case the constants C will be controlled by some negative power of h. This will say in this case, that if ω_1 is shown to be exponentially small, then one should expect that the best α_1 has necessarily the same property. We shall come back to this point in Chapter 17.

6.5 Fokker-Planck Operators and Kinetic Equations

If one considers the evolution of probability measures in the phase space, which is the point of view of kinetic equations, the Fokker-Planck equation is written in the form

$$\begin{cases} \partial_t f + v \cdot \partial_x f - \partial_x V(x) \cdot \partial_v f - \partial_v \cdot (\partial_v + v) f = 0 \\ f(t=0) = f_0 \end{cases}$$

which means that one considers the operator $K_1 = M^{1/2} K M^{-1/2}$. On the contrary probabilists are sometimes interested in the dual equation which governs the evolution of observables

$$\begin{cases} \partial_t g + v \cdot \partial_x g - \partial_x V(x) \cdot \partial_v g + (-\partial_v + v) \cdot \partial_v g = 0 \\ g(t=0) = g_0 \end{cases}$$

and the operator is $K_0 = M^{-1/2} K M^{1/2}$. We consider the case when the normalized Maxwellian $\mu = \left(\int M \right)^{-1} M$ is a probability measure. Theorem 6.4 says that for any $g_0 \in M^{-1/2} L^2(\mathbb{R}^{2n})$ (and this can be extended to $g_0 \in M^{-1/2} \mathcal{S}'(\mathbb{R}^{2n})$) we have

$$\mu \left(|g(t) - \mu(g_0)|^2 \right) \leq C^2 e^{-2\alpha_1 t} \ .$$

This permits exponentially increasing initial data but does not give any L^∞ bound on $g(t)$. On the other side, for any $f_0 \in M^{1/2} L^2(\mathbb{R}^{2n})$ (and this can be extended to $f_0 \in M^{1/2} \mathcal{S}'(\mathbb{R}^{2n})$), Theorem 6.4 gives

$$\int_{\mathbb{R}^{2n}} \left| f(t) - \left(\int_{\mathbb{R}^{2n}} f_0 \right) \mu \right|^2 M^{-1} \, dx \, dv \leq C^2 e^{-2\alpha_1 t} \ ,$$

which gives a strong control of $\left| f(x, v, ; t) - \left(\int f_0 \right) \mu(x, v) \right|$ but does not hold for general $f_0 \in L^1(\mathbb{R}^{2n})$. From this point of view and if one wants to treat with such techniques nonlinear kinetic equations, Theorem 6.4 is not completely satisfactory because the preservation of the L^1-norm is one of the few preserved quantities which are used for nonlinear problems. With this respect, it would be interesting to extend the proof of exponential decay for initial data $f_0 \in L^1(\mathbb{R}^{2n})$.

7

Hypoellipticity and Nilpotent Groups

7.1 Introduction

The analysis of left invariant homogeneous operators on a nilpotent graded Lie group has played a very important role in the understanding of the hypoellipticity. There are two important points:

- There is a beautiful characterization of the hypoellipticity for these operators initially conjectured by C. Rockland [Roc]. We will describe these results in Section 7.4.
- These models appear as the right approximation (in a suitable sense) of many more general (and non necessarily invariant) differential operators. We will describe the result in Chapter 8.

7.2 Nilpotent Lie Algebras

We refer to [No5] and [HelNo3] for a more comprehensive description of the theory. Let us explain this roughly. The starting point is an abstract Lie algebra $\mathcal{G}$ (think of a subalgebra of the up-triangular maps) admitting the following decomposition:

$$\mathcal{G} = \oplus_{j=1}^{r} \mathcal{G}_j \, ,$$

with the property that the $\mathcal{G}_j$'s are vector subspaces in direct sum satisfying

$$[\mathcal{G}_i, \mathcal{G}_j] \subset \mathcal{G}_{i+j} \, .$$

Usually, we will consider the case when $\mathcal{G}$ is generated by $\mathcal{G}_1$, that is the case of a **stratified algebra** but it could be necessary to consider other cases in order to treat the Hörmander operators of type 2.

Definition 7.1.
We will say that a graded Lie algebra is stratified if it is generated by $\mathcal{G}_1$. We will say that a graded Lie algebra is stratified of type 2 if it is generated by $\mathcal{G}_1 \oplus \mathcal{G}_2$.

The second case occurs for the type 2 Hörmander's operator, where X_0 is considered in $\mathcal{G}_2$ and the Lie algebra is generated by $\mathcal{G}_1$ and X_0.

In the case of these algebras there exists a global diffeomorphism from $\mathcal{G}$ onto a simply connected group G via the exponential map (restriction of the exponential map on matrices). This gives a group structure on $\mathcal{G}$ denoted by "$\circ$" (which is then identified with G). Typically when the algebra is of rank 2, we get

$$a \circ b = a + b + \frac{1}{2}[a, b] \ .$$

When the algebra is of rank 3, we have:

$$a \circ b = a + b + \frac{1}{2}[a, b] + \frac{1}{12}[a, [a, b]] + \frac{1}{12}[[a, b], b] \ .$$

One can then identify an element v of $\mathcal{G}$ to a left invariant vector field $\rho(v)$ on the group G (identified with $\mathcal{G}$ via the exponential map) by the formula:

$$(\rho(v)g)(u) = \frac{d}{ds}g(u \circ (sv))_{|s=0} \ . \tag{7.1}$$

In the same way, an element of the enveloping algebra $\mathcal{U}(\mathcal{G})$, that is a non-commutative polynomial

$$P := \sum a_\alpha Y^\alpha \ ,$$

(where Y_ℓ is a basis of $\mathcal{G}$, or of $\mathcal{G}_1$ in the stratified case, and $a_\alpha \in \mathbb{C}$) can be identified to a left invariant operator on the group G.

On this Lie algebra, we have a natural δ_t which is linear map with the property that

$$\delta_t(X) = t^j X \ , \quad \text{if } X \in \mathcal{G}_j \ . \tag{7.2}$$

It is then immediate to extend δ_t to $\mathcal{U}(\mathcal{G})$ and to define what is an homogeneous operator of order m in the enveloping algebra and we denote by $\mathcal{U}_m(\mathcal{G})$ the linear space of these operators. We note that the Hörmander operator $\sum_j Y_j^2$ where Y_j is a basis of $\mathcal{G}_1$ is in $\mathcal{U}_2(\mathcal{G})$.

7.3 Representation Theory

We now explain very briefly the Kirillov theory. Starting from a unitary representation of the group G, we can always attach a representation of the Lie algebra . Among these representations, the irreducible representations will play an important role. The Kirillov theory permits to associate to any element of the dual $\mathcal{G}^*$ of $\mathcal{G}$ an irreducible representation. Moreover this theory says that any unitary representation can be represented in this way. Finally two irreducible representations are unitarily equivalent if they belong to the same orbit.

In order to be more precise, let us give the definition of induced representation (which is due to Mackey). We give here a rather "pedestrian" definition. The starting point is a subalgebra $\mathcal{H} \subset \mathcal{G}$ and a linear form on $\mathcal{G}$ such that: $\ell([\mathcal{H}, \mathcal{H}]) = 0$. Then we will associate a representation $\pi_{\ell, \mathcal{H}}$ of the group G into $L^2(\mathbb{R}^{k(\ell)})$, which is uniquely defined modulo a unitary conjugation, where $k(\ell)$ is the codimension of $\mathcal{H}$ in $\mathcal{G}$. For this construction and using the nilpotent character, we can find $k = k(\ell)$ independent vectors $e_1, \ldots, e_k$ such that any g can be written in the form

$$g = h \exp s_k e_k \, \exp s_{k-1} e_{k-1} \cdots \exp s_1 e_1 \;, \tag{7.3}$$

and if

$$\mathcal{A}_j = \mathcal{H} \oplus \mathbb{R} e_k \oplus \cdots \oplus \mathbb{R} e_{k-j+1} \;, \tag{7.4}$$

then $\mathcal{A}_{j-1}$ is an ideal of codimension 1 in $\mathcal{A}_j$.

With this construction, one can get that $g \mapsto (s, h)$ is a global diffeomorphism from $\mathcal{G}$ onto $\mathbb{R}^k \times \mathcal{H}$.

The induced representation is given by:

$$(\pi_{\ell, \mathcal{H}}(\exp a)f)(t) = \exp i\langle \ell, h(t, a)\rangle \; f(\sigma(t, a)) \;.$$

Here $h(t, a)$ and $\sigma(t, a)$ are defined by:

$$\exp t_k e_k \cdots \exp t_1 e_1 \, \exp a$$
$$= \exp h(t, a) \, \exp \sigma_k(t, a)e_k \cdots \exp \sigma_1(t, a)e_1 \;.$$

The corresponding representation of $\mathcal{G}$ is defined by:

$$(\pi_{\ell, \mathcal{H}}(a)f)(t) = \frac{d}{ds}((\pi_{\ell, \mathcal{H}}(\exp sa)f)(t))_{s=0} \;,$$

with $h(t, a) \in \mathcal{H}$, according to Formula (7.3). More explicitly, we get:

$$\pi_{\ell, \mathcal{H}}(a) = i\langle \ell, h'(t, a)\rangle + \sum_{j=1}^{k} \sigma'_j(t, a)\partial_{t_j} \;, \tag{7.5}$$

where

$$h'(t, a) = \frac{d}{ds}(h(t, sa))_{/s=0} \;, \; \sigma'(t, a) = \frac{d}{ds}(\sigma(t, sa))_{/s=0} \;.$$

There are two particular cases, which are interesting. When $\ell = 0$, we get the standard extension of the trivial representation of the subgroup H of G. It can be considered as a representation on $L^2(G/H)$. An interesting problem is to characterize the maximal hypoellipticity of $\pi_{0, \mathcal{H}}(P)$ for P in $\mathcal{U}_m(\mathcal{G})$ (elements of $\mathcal{U}(\mathcal{G})$ with degree m).

The second point is when $\mathcal{H}$ is of maximal dimension, for a given ℓ, with the above property. In this case, one can show that the representation is irreducible.

By this way one can construct all the irreducible representations. Starting this time from an ℓ in $\mathcal{G}^*$, one can construct a maximal subalgebra V_ℓ such that $\ell([V_\ell, V_\ell]) = 0$. One can show that the codimension $k(\ell)$ of V_ℓ is equal to $\frac{1}{2}$ rank B_ℓ where B_ℓ is the two-form:

$$\mathcal{G} \times \mathcal{G} \ni (X, Y) \mapsto \ell([X, Y]) .$$

For $a \in \mathcal{G}$, we define by $(\operatorname{ad} a)^*$ the adjoint of $\operatorname{ad} a$:

$$(\operatorname{ad} a)^* \ell(b) = \ell([a, b]) .$$

The group G acts naturally on $\mathcal{G}^*$ by:

$$g \mapsto g\ell = \sum_{k=0}^{r} \frac{1}{k!} (\operatorname{ad} - a)^{*^k} \ell ,$$

with $g = \exp a$. This action is called the coadjoint action.

So what we know from Kirillov's theory is that if ℓ and $\tilde{\ell}$ are on the same orbit, then the corresponding unitary representations are equivalent.

Conversely two different orbits give two non equivalent irreducible representations, so, one can identify the set $\hat{G}$ of the irreducible representations of G with the set of the G-orbits in $\mathcal{G}^*$:

$$\hat{G} = \mathcal{G}^*/G . \tag{7.6}$$

7.4 Rockland's Conjecture

The following theorem was conjectured by C. Rockland [Roc] and proved in full generality by B. Helffer and J. Nourrigat [HelNo1, HelNo2] (See later [Glo1, Glo2, Mel]):

Theorem 7.2.
An element P in $\mathcal{U}_m(\mathcal{G})$ is hypoelliptic if and only if, for any non trivial representation π of $\mathcal{G}$, $\pi(P)$ is injective in $\mathcal{S}_\pi$ (defined below in (7.10)).

For the basic example, the proof of the hypoellipticity of $\sum_{j=1}^{p} Y_j^2$ as a consequence of Theorem 7.2 becomes trivial. The condition that u satisfies $\pi(P)u = 0$ implies $\pi(Y_j)u = 0$, for any $j = 1, \ldots, p$. This implies, in the stratified case that $\pi(Y)u = 0$ for any $Y \in \mathcal{G}$. A characterization of the irreducibility of π and the assumption that π is not trivial gives $u = 0$.

An important ingredient is the proof of maximal inequalities of the type

$$||\pi(Q)u|| \leq C_{Q,\pi} ||\pi(P)u|| , \quad \forall u \in \mathcal{S}_\pi \tag{7.7}$$

for all $Q \in \mathcal{U}_m(\mathcal{G})$ and all irreducible representations π of G. Here $\mathcal{S}_\pi$ is the space of C^∞ vectors of the representation (see (7.10)).

The goal is then to deduce, from this family of inequalities, the corresponding maximal inequality for P, that is

$$||Qu|| \leq C_Q \, ||Pu|| \, , \, \forall u \in \mathcal{S}(G) \, , \tag{7.8}$$

for all $Q \in \mathcal{U}_m(\mathcal{G})$.

A fundamental point for proving these maximal inequalities, , which will imply hypoellipticity, is to show that the constant $C_{Q,\pi}$ can be chosen independent of π and to have it for Q belonging to a basis of $\mathcal{U}_m(\mathcal{G})$. We can not explain here all the recursion argument, which is strongly related to Kirillov's theory, but we would like to emphasize on some ingredients of the proof, which can give an interesting light for other problems.

Remark 7.3.

We note for future use, that in this case we get also a maximal inequality for operators $\pi(P)$ associated to other, non necessarily irreducible, representations π. For example, this can be applied to the operator $\pi_{\ell,\mathcal{H}}(P)$, where $\pi_{\ell,\mathcal{H}}$ is the representation introduced in Section 7.3.

7.5 Spectral Properties

We assume for simplicity that we are in the stratified case. Sometimes, we shall consider the stratified case of type 2 (in this case the operators will be assumed to be homogeneous of even order). For $m \in \mathbb{N}$ and for a unitary representation in $\mathcal{H}_\pi$, the Sobolev spaces H_π^m are naturally defined. H_π^0 is by definition the space of representations $\mathcal{H}_\pi$ (It is enough for our applications to consider[1] the case when $\mathcal{H}_\pi := L^2(\mathbb{R}^k)$ for some $k \geq 0$). Then:

$$H_\pi^m = \{u \in H_\pi^0 \mid (\pi(Y))^\alpha u \in \mathcal{H}_\pi\} \, . \tag{7.9}$$

These spaces have a natural Hilbert structure. One can also define the H_π^m for $m \in \mathbb{Z}$ (and by complex interpolation for $m \in \mathbb{R}$). Moreover

$$\mathcal{S}_\pi = \cap_m H_\pi^m \, , \tag{7.10}$$

is dense in each of the H_π^m. When π is irreducible, then we have compact imbedding of $H_\pi^{m'}$ into H_π^m if $m' > m$, and we can identify $\mathcal{S}_\pi$ and $\mathcal{S}(\mathbb{R}^k)$ and $\mathcal{S}'(\mathbb{R}^k)$ with

$$\mathcal{S}_\pi' := \cup_m H_\pi^m \, . \tag{7.11}$$

If P satisfies the Rockland condition for all the degenerate π's, that is for irreducible representations associated to elements $\ell \in \mathcal{G}^*$ such that $\ell_r = \ell|_{\mathcal{G}_r} = 0$, then $\pi(P)$ is a Fredholm operator of H_π^m onto $H_\pi^0 = \mathcal{H}_\pi$. If P is symmetric, $\pi(P)$ is essentially self-adjoint on $\mathcal{S}_\pi$, with domain H_π^m, and has consequently

---

[1] We have in particular seen that the irreducible representations are of this type.

a compact resolvent, when $m > 0$.

One of the steps in the proof is to show an inequality with remainder:

$$||\pi(Q)u|| \leq C_{Q,\pi}(||\pi(P)u|| + ||u||) \,, \ \forall u \in \mathcal{S}_\pi \,. \tag{7.12}$$

There are actually many examples in physics which can be recognized as $\pi(P)$ for some homogeneous $P \in \mathcal{U}(\mathcal{G})$. The most standard is the harmonic oscillator, which can be recognized as a $\pi(Y_1^2 + Y_2^2)$ in the case when $\mathcal{G}$ is the three dimensional Heisenberg algebra and π is the irreducible representation attached to $\ell_2([Y_1, Y_2]) = 1$.

But let us also look at $Y_1^2 + Y_2^2 - i[Y_1, Y_2]$. This operator is not hypoelliptic. But it satisfies the degenerate Rockland condition. The corresponding $\pi(P)$ is $-\partial_t^2 + t^2 - 1$. It has effectively a compact resolvent but is not injective.

The important remark is that if we want to prove by this approach that some operators have compact resolvent it is enough to consider the so called degenerate Rockland's condition.

8

Maximal Hypoellipticity for Polynomial of Vector Fields and Spectral Byproducts

8.1 Introduction

We assume that we are given p vector fields on an open set $\Omega \subset \mathbb{R}^n$, satisfying the rank r Hörmander condition (see Section 2.1). Let $\mathcal{P}$ be a non-commutative polynomial of degree less or equal to m of vector fields with C^∞ coefficients, that is an operator of the form

$$\mathcal{P} := \sum_{|\alpha| \leq m} a_\alpha(x) X^\alpha ,$$

where $\alpha \in \{1, \ldots, p\}^k$, $|\alpha| = k$.

We say that $\mathcal{P}$ is maximally hypoelliptic if, for any compact $K \subset \Omega$, there exists a constant $C_K > 0$ such that:

$$\sum_{|\alpha| \leq m} \|X^\alpha u\|^2 \leq C \left(\|\mathcal{P}u\|_0^2 + \|u\|_0^2 \right) , \tag{8.1}$$

$$\forall u \in C_0^\infty(\Omega) \text{ s.t. } \operatorname{supp} u \subset K .$$

Localized version:

We say that the property of maximal hypoellipticity is satisfied at x_0 if there exists an open neighborhood of x_0, in which the previous property is satisfied with Ω replaced by ω_{x_0}.

One can show[1] that this inequality (jointly with the Hörmander condition) implies the hypoellipticity of $\mathcal{P}$ in Ω. Using a technique of lifting of the problem on a nilpotent universal group and fine harmonic analysis, Rothschild-Stein [RoSt] have proven the following theorem.

Theorem 8.1.
If the vector fields X_j ($j = 1, \ldots, p$) satisfy the Hörmander condition at any point of Ω, then the operator $L_1 := \sum_{j=1}^p X_j^2$ is maximally hypoelliptic.

---

[1] A standard reference is [Tr1]

Similarly, they treat also the case of the Hörmander's operator of type 2.

Theorem 8.2.
If the vector fields X_0, X_j ($j = 1, \ldots, p$) satisfy the Hörmander condition at any point of Ω, then the operator $L_2 := \sum_{j=1}^{p} X_j^2 + X_0$ is maximally hypoelliptic in the following sense. For any compact $K \subset \Omega$, there exists a constant $C_K > 0$ such that:

$$||X_0 u||^2 + \sum_{j=1}^{p} ||X_j u||^2 + \sum_{j,k=1}^{p} ||X_j X_k u||^2 \leq C \left(||L_2 u||_0^2 + ||u||_0^2 \right) , \tag{8.2}$$

$$\forall u \in C_0^\infty(\Omega) \ s.t. \ \operatorname{supp} u \subset K .$$

We observe that in ((8.2)) we have given the weight 2 to X_0 and the weight 1 to the other vector fields X_j. With this convention, L_2 appears to be an homogeneous operator of degree 2.

Because this property of maximal hypoellipticity is much stronger and robust than hypoellipticity, there is a hope to characterize the maximally hypoelliptic operators (cf [HelNo3]). We have already seen that all the homogeneous left invariant hypoelliptic operators are maximally hypoelliptic. Our aim in this chapter is to give a flavor about the nature of the criteria and also to explain some byproducts of the proofs.

These maximally hypoelliptic operators are more stable in the sense that this property depends only on the "principal part"

$$\mathcal{P}^0 := \sum_{|\alpha|=m} a_\alpha(x) X^\alpha .$$

8.2 Rothschild-Stein Lifting and Towards a General Criterion

L.P. Rothschild and E. Stein [RoSt] (see also Goodman [Goo]) have shown that the analysis of the maximal hypoellipticity of the operator $\sum_j X_j^2$ or of $X_0 + \sum_j X_j^2$ can be deduced, when the vector fields satisfy the rank r condition on an open set Ω, from the analysis of a corresponding invariant homogeneous operator defined on a free nilpotent Lie group of rank r. The free Lie algebra of rank r with p generators $\mathcal{G} = \mathcal{G}^{r,p}$ is the maximal Lie algebra with this property, that is the only relations existing are the necessary conditions satisfied by any Lie algebra of rank r with p generators. For example, the rank 2 free algebra with p generators has a basis:

$$Y_1, \cdots, Y_p ; [Y_i, Y_j] \ (i < j) .$$

One can show that there exists a unique linear application λ from $\mathcal{G}$ into the algebra of the vector fields defined on Ω such that:

$$\lambda(Y_i) = X_i \ .$$

Moreover λ is a partial homomorphism of rank r, that is:

$$[\lambda(a_i), \lambda(a_j)] = \lambda([a_i, a_j]) \ , \ \forall a_i \in \mathcal{G}_i, a_j \in \mathcal{G}_j \text{ with } i + j \leq r \ .$$

For $x \in \Omega$, we denote by λ_x the map from $\mathcal{G}$ into $T_x\Omega$ defined by:

$$\lambda_x(a) = (\lambda(a))_x \ , \ \forall a \in \mathcal{G} \ ,$$

and by λ_x^* the transposed map from $T_x^*\Omega$ into $\mathcal{G}^*$. For any $u \in \mathcal{G}$, we can write:

$$\lambda(u) = \sum_j a_j(x, u) \partial_{x_j} \ ,$$

and we denote by $y(t, u, x)$ the integral curve of $\lambda(u)$ starting from x at $t = 0$. We then introduce:

$$(x, u) \mapsto \psi(x, u) := y(1, u, x) \ ,$$

which is a C^∞ map in a neighborhood of say $(x_0, 0)$ in $\Omega \times \mathcal{G}$ with value in Ω. For any $x \in \Omega$, we can then introduce, for $f \in C^\infty(\Omega)$,

$$(W_x f)(u) = f(\psi(x, u)) \ .$$

Then we have the following approximation theorem

$$\lim_{t \to 0} \delta_t^{-1} W_x \lambda(\delta_t v) W_x^{-1} = \rho(v) \ ,$$

where $\rho(v)$ is the left invariant vector field associated with v, by (7.1). In particular, the Rothschild-Stein theory permits to deduce that if $P_m(Y_1, \cdots, Y_p)$ is hypoelliptic then $P(X_1, \cdots, X_p)$ is maximally hypoelliptic. This theorem is very strong at the level of the regularity, but still too weak in the sense that it gives only a sufficient condition of maximal hypoellipticity, which can be quite far from necessary. The other weakness is that it does not give a way to find criteria for the hypoellipticity of $P_m(Y_1, \cdots, Y_p)$.

So it is natural to formulate after [HelNo3] the following conjecture.

Conjecture 8.3.
There exists a subset Γ of $\mathcal{G}^*$, which is

- closed,
- stable by the coadjoint action,
- and stable by dilation,

such that $\mathcal{P}$ is maximally hypoelliptic at x_0, if and only if $\pi_\ell(P)$ is injective for any $\ell \in \Gamma \setminus \{0\}$.

The candidate $\Gamma = \Gamma_{x_0}$ can be defined as follows:

Definition 8.4.
The set Γ_{x_0} is attached to the vector fields $X_1, \cdots, X_p$ by:

$$\Gamma_{x_0} = \{\ell \in \mathcal{G}^* \mid \ell = \lim_{n \to +\infty} \delta_{t_n}^* \ell_n \mid$$
$$\text{for some sequence } (t_n, \ell_n) \text{ with } t_n > 0 , \ t_n \to 0 , \ \ell_n = \ell_{x_n, \xi_n}\} . \tag{8.3}$$

Here $\ell_{x_n, \xi_n} \in \mathcal{G}^*$ is defined by

$$\ell_{x_n, \xi_n}(Y) = \frac{1}{i}\sigma(\lambda(Y))(x_n, \xi_n) , \ \forall Y \in \mathcal{G} ,$$

where $\sigma(\lambda(Y))$ denotes the symbol of $\lambda(Y)$. Moreover, we impose that

$$|\xi_n| \to +\infty \text{ and } x_n \to x_0 .$$

Note that

$$\frac{1}{i}\sigma(\lambda(Y))(x_n, \xi_n) = \lambda_{x_n}^* \xi_n(Y) ,$$

so

$$\ell_{x_n, \xi_n} = \lambda_{x_n}^* \xi_n .$$

It can be shown that Γ_{x_0} is a closed G-invariant homogeneous set in $\mathcal{G}^*$.

The necessity of the conjecture with $\Gamma = \Gamma_{x_0}$ was proved by J. Nourrigat (cf Chapter 3 in [HelNo3]) and a microlocal version for systems is given in [No3]. The sufficiency is proved in many cases [HelNo3], containing in particular the case when $[\mathcal{G}^2, \mathcal{G}^2] = 0$ (this condition will be always satisfied in our examples). There is a general proof for systems given by J. Nourrigat in the 80's [No6, No7]. We will discuss some aspects of the theory in Chapter 10.

A particular case is the case when $\mathcal{P} = \pi_{0, \mathcal{H}}(P)$. In this case Γ is the so called "spectrum of the induced representation" or "support of the induced representation":

$$\Gamma = \mathrm{Sp}\,(\pi_{0, \mathcal{H}}) = \overline{G \cdot \mathcal{H}^\perp} \tag{8.4}$$

There is a microlocal version of the conjecture. First recall that, if $u \in \mathcal{D}'(\Omega)$ and if (x_0, ξ_0) is a point in $T^*\Omega \setminus \{0\}$, one says that this point does not belong to the Wave Front of u (denoted by $WF\,u$) if for a cut-off function $\chi \in C_0^\infty(\Omega)$ such that $\chi(x_0) \neq 0$ there exists a conical neighborhood of ξ_0 in which $\widehat{\chi u}(\xi)$ is rapidly decreasing. This defines a closed subset in $T^*\Omega \setminus \{0\}$ whose projection on Ω gives the so called singular support of u: Sing supp u. Then a microlocally hypoelliptic operator is an operator such that

$$WF(u) \subset WF(Pu) , \ \forall u \in \mathcal{D}'(\Omega) . \tag{8.5}$$

More generally, this notion can be localized[2] in cones $\Gamma^* := \omega \times \gamma$ of $T^*\Omega \setminus \{0\}$ or at the point (x_0, ξ_0) in the sense that there exists an open conic neighborhood Γ_{x_0, ξ_0}^* of (x_0, ξ_0) in which the operator is microlocally hypoelliptic.

---

[2] Just write

$$WF(u) \cap \Gamma^* \subset WF(Pu) , \ \forall u \in \mathcal{D}'(\Omega) .$$

Similarly (we refer to [HelNo3] for a precise definition or to Chapter 10 in a particular case) we can introduce the notion of maximal microhypoellipticity and Helffer and Nourrigat [HelNo3] formulate the following conjecture.

Conjecture 8.5.
There exists a subset Γ of $\mathcal{G}^*$, which is closed, stable by the coadjoint action and by dilation, such that $\mathcal{P}$ is microlocally maximally hypoelliptic at (x_0, ξ_0) if and only if $\pi_\ell(P)$ is injective for any $\ell \in \Gamma \setminus \{0\}$.

The set Γ_{x_0} appearing in the "local" conjecture is then replaced at a point $(x_0, \xi_0) \in \Omega \times \mathbb{R}^d \setminus \{0\}$ by:

Definition 8.6.
The set Γ_{x_0, ξ_0} is defined by

$$\Gamma_{x_0, \xi_0} = \{\ell \in \mathcal{G}^* \mid \ell = \lim_{n \to +\infty} \delta_{t_n}^* \ell_n \mid$$
$$\text{for some sequence } (t_n, \ell_n) \text{ with } t_n > 0 \, , \, t_n \to 0 \, , \, \ell_n = \ell_{x_n, \xi_n}\} \, , \tag{8.6}$$

with in addition

$$\lim_{n \to +\infty} x_n = x_0 \, , \quad \lim_{n \to +\infty} \frac{\xi_n}{|\xi_n|} = \frac{\xi_0}{|\xi_0|} \, .$$

This set is also closed, G-stable and dilation invariant. Note also that the definition of Γ_{x_0, ξ_0} only depends on $\xi_0/|\xi_0|$, so: $\Gamma_{x_0, t\xi_0} = \Gamma_{x_0, \xi_0}$.

8.3 Folland's Result

We refer here to [Fol] and to the discussion in [HelNo3] (p. 27-28). Let V be a real vector space admitting a decomposition as a direct sum of spaces V_i $(1 \leq i \leq r)$. For every $t > 0$, we define the dilation h_t on V by

$$h_t\left(\sum_{j=1}^r x_j\right) = \sum_{j=1}^r t^j x_j \, , \quad \text{for } x_j \in V_j \, . \tag{8.7}$$

We say that a differential operator P on V is homogeneous of degree m if

$$P(f \circ h_t) = t^m (Pf) \circ h_t \, , \quad \forall f \in C^\infty(V) \, . \tag{8.8}$$

Let $X_1, \cdots, X_p$ a system of C^∞ real vector fields homogeneous of degree 1 and satisfying the rank r Hörmander condition.
Let $\mathcal{G}$ be the free nilpotent Lie Algebra of rank r with p generators. Let $L(V)$ the Lie algebra of C^∞ vector fields on V. Then there exists a unique linear map from $\mathcal{G}$ into $L(V)$ such that:

$$\lambda(Y_j) = X_j \, ,$$
$$\lambda([a_j, a_k]) = [\lambda(a_j), \lambda(a_k)] \, , \quad \text{for } a_j \in \mathcal{G}_j, a_k \in \mathcal{G}_k \, . \tag{8.9}$$

One can verify that λ is an homomorphism (compare with the general Rothschild-Stein theory where we have only a partial homomorphism of rank r). We observe indeed that an homogeneous vector field of degree $> r$ is necessarily identically 0.

We now define $\mathcal{H}$ as the subspace of $\mathcal{G}$ generated by the a's such that $\lambda(a)$ vanishes at 0. It is immediate to see that $\mathcal{H}$ is a subalgebra of $\mathcal{G}$ and stable by the dilations of G:

$$\lambda \circ \delta_t = h_t \circ \lambda . \tag{8.10}$$

Moreover the codimension of $\mathcal{H}$ is equal to $\dim V$. Folland [Fol] has proved the

Proposition 8.7.
There exists a global diffeomorphism θ from V onto $\mathbb{R}^{\dim V}$ with Jacobian 1, such that for all $a \in \mathcal{G}$ and $f \in C_0^\infty(\mathbb{R}^{\dim V})$, we have:

$$\pi_{0,\mathcal{H}}(a)f = (\lambda(a)(f \circ \theta)) \circ \theta^{-1} . \tag{8.11}$$

Moreover, the map $f \mapsto f \circ \theta$ sends $\mathcal{S}(\mathbb{R}^{\dim V})$ onto $\mathcal{S}(V)$.

The Witten situation

Let us consider the Witten case. We take an homogeneous polynomial Φ of degree r on $\mathbb{R}^n$. We take $V_1 = \mathbb{R}_x^n$, $V_2 = \cdots = V_{r-1} = \{0\}$, $V_r = \mathbb{R}_t$. We define:

$$X_j = \partial_{x_j} , \quad X_{j+n} = (\partial_{x_j}\Phi)\partial_t .$$

We assume that $r \geq 2$. Then it is clear that $\mathcal{H}$ in $\mathcal{G}$, is generated by $\mathcal{G}_1'' \oplus \mathcal{G}_2 \oplus \cdots \mathcal{G}_{r-1}$ where $\mathcal{G}_1''$ is generated by the X_{j+n} $(j = 1, \ldots, n)$.

What is the spectrum of $\pi_{0,\mathcal{H}}$? Coming back to the definition (8.4), we obtain that ℓ in $\mathcal{G}^*$ is in the spectrum of the representation, if there exists a sequence (x_n, τ_n) such that:

$$
\begin{aligned}
\ell_1(X_{j+n}) &= \lim_{n\to+\infty}(\partial_{x_j}\Phi)(x_n)\tau_n , && \text{for } j = 1, \ldots, n , \\
\ell_2([X_k, X_{j+n}]) &= \lim_{n\to+\infty}(\partial_{x_k}\partial_{x_j}\Phi)(x_n)\tau_n , && \text{for } j, k = 1, \ldots, n , \\
\ell_2([X_{k+n}, X_{j+n}]) &= 0 , && \text{for } j, k = 1, \ldots, n , \\
\cdots & \qquad \cdots \\
\ell_q(\mathrm{ad}\, X^\alpha X_{\ell+n}) &= \lim_{n\to+\infty}(\partial_x^\alpha \partial_{x_\ell}\Phi)(x_n)\tau_n , && \text{for } |\alpha| + 1 = q \leq r .
\end{aligned}
$$

Note that the Hörmander condition is satisfied if the polynomial Φ is not identically constant. Note also that the representation π_τ is irreducible for $\tau \neq 0$, if $\lim_{x\to\infty}\sum_{|\alpha|\geq 1}|\partial_x^\alpha\Phi(x)| = +\infty$.

Remark 8.8.
It is sometimes better to work directly with a smaller algebra than the free algebra, by choosing an algebra taking into account the bracket properties of the vector fields X_j (for example if two vector fields commute).

If for the hypoellipticity of $\pi_{0,\mathcal{H}}(P)$, it is the injectivity of $\pi(P)$ for any π in the spectrum of $\pi_{0,\mathcal{H}}$. The property that $\pi_{\tau}(P)$ has a compact resolvent for $\tau \neq 0$ can be obtained on the basis that $\pi(P)$ is injective for all π in the spectrum of $\pi_{0,\mathcal{H}}$ which are degenerate on $\mathcal{G}_r$ and non trivial.

The investigation of the proof in Helffer-Nourrigat [HelNo3] permits to separate the case $\tau > 0$ and the case $\tau < 0$. This corresponds to a disjoint analysis of the microlocal hypoellipticity at the two points in $T_0^* \mathbb{R}^{n+1} \setminus \{0\}$, where the operator $\pi_{(0,\mathcal{H})}$ is not elliptic.

Let us treat the easy case. This is the case when Φ is elliptic outside 0 or more generally when $\nabla \Phi$ does not vanish on the unit sphere $\mathbb{S}^{n-1}$. In this case, we get easily that the only representations which belong to this "degenerate" spectrum are the representations which corresponds in Kirillov's Theory to elements of $\mathcal{G}^* \cap (\mathcal{G}^2)^{\perp}$, i.e. which vanish on $\mathcal{G}^2 = \sum_{j=2}^{r} \mathcal{G}_j$. One gets immediately that for these π's (π not trivial), $\pi(P)$ is injective.

Of course this result can be proved quite easily by the criterion given in Proposition 3.1.

When Φ is not elliptic, there are other cases where one can give an answer but we postpone this to the next chapters. The simplest example is $\Phi(x_1, x_2) = \pm x_1^2 x_2^2$. But this case can again be treated (for the sign $-$) by Proposition 3.1.

8.4 Discussion on Rothschild-Stein and Helffer-Métivier-Nourrigat Results

We would like to analyze the properties of

$$L := \Sigma_{j=1}^{p_1} Y_j^2 + \frac{i}{2} \sum_{j,k} b_{jk} [Y_j, Y_k] , \qquad (8.12)$$

where b_{jk} is a real antisymmetric matrix and the Y_j's form a basis of $\mathcal{G}_1$ and $\mathcal{G}_1$ generates a stratified Lie algebra:

$$\mathcal{G} = \oplus_{j=1}^{r} \mathcal{G}_j .$$

Let us introduce some definitions. If ρ is a real antisymmetric matrix, we can define its trace norm by

$$\|\rho\|_1 = \sum_j |\rho_j| , \qquad (8.13)$$

where the ρ_j are the eigenvalues of ρ.

We denote by S the subspace of the real antisymmetric matrices such that

$$\sum_{j,k} s_{jk} [Y_j, Y_k] = 0 . \qquad (8.14)$$

Theorem 8.9.
Suppose that $\mathcal{G}$ is not an Heisenberg algebra. Then L is hypoelliptic if and only if:

$$\sup_{\|\rho\|_1 \leq 1 \,,\, \rho \in S^\perp} |\operatorname{Tr}(b\rho)| < 1 \,. \tag{8.15}$$

(Here $S^\perp$ is the set of ρ's such that $\operatorname{Tr}(s\rho) = 0$, $\forall s \in S$.)

The sufficiency part is proved in [RoSt] and the assumption that $\mathcal{G}$ is not an Heisenberg algebra is not needed. The case of an Heisenberg algebra can be treated separately and discrete phenomena appear. The result was completed in [Hel1] (or[3] [Hel2])

For the comparison with Witten Laplacians. It is enough to concentrate on the following case. The space $\mathcal{G}_1$ admits a decomposition:

$$\mathcal{G}_1 = \mathcal{G}_1' \oplus \mathcal{G}_1'' \,, \tag{8.16}$$

with

$$\dim \mathcal{G}_1' = \dim \mathcal{G}_1'' \,.$$

Moreover

$$[\mathcal{G}^2, \mathcal{G}^2] = 0 \,,$$

and

$$\dim \mathcal{G}_r = 1 \,.$$

Our operator is

$$L := \sum_{j=1}^{p_1'} (Y_j')^2 + \sum_{j=1}^{p_1'} (Y_j'')^2 + i \sum_{j=1}^{p_1'} [Y_j', Y_j''] \,, \tag{8.17}$$

where Y_j' ($j = 1, \ldots, p_1'$) (resp. Y_j'' ($j = 1, \ldots, p_1'$)) denotes a basis of $\mathcal{G}_1'$ (resp. $\mathcal{G}_2^j$).

Actually, we are not exactly interested in this operator but more in the properties of $\Pi(P)$, where Π is some representation of the enveloping algebra. This representation Π will actually be an induced representation $\pi_{(0,\mathcal{H})}$, with $\mathcal{H} = \mathcal{G}_1'' \oplus \mathcal{G}_2 \oplus \cdots \oplus \mathcal{G}_{r-1}$.

---

[3] Note that Lemma 3.2 as stated in these notes is obviously wrong and that the argument has to be corrected, like in the proof of (5.10) in [Hel3].

Remark 8.10.

In a completely different context[4], [GHH] analyze the properties of some "Dirac type" operator presenting similar "bad" properties. More precisely they consider on $L^2(\mathbb{R}^2, \mathbb{C}^2)$ the operator

$$(-\partial_x^2 - \partial_y^2 + x^2 y^2) \otimes I + x\sigma_3 + y\sigma_1 ,$$

where the σ_j's are the Pauli matrices (see Remark 3.11). Here $\sigma_3 = \begin{pmatrix} 1 & 0 \\ 0 & -1 \end{pmatrix}$.

This is in some sense a "vector valued" example of the type described in (8.12), if one observes that:

$$[\partial_x, xy] = y , \quad [\partial_y, xy] = x .$$

---

[4] We thank J. Hoppe for this remark

9

On Fokker-Planck Operators
and Nilpotent Techniques

9.1 Is There a Lie Algebra Approach
for the Fokker-Planck Equation?

We give here some remarks about the possible Lie algebra structures which can be associated with the Fokker-Planck equation.

First approach:

As suggested in the introduction one can write the Fokker-Planck operator as

$$K = X_0 - \sum_\ell \left((X_1^\ell)^2 + (X_2^\ell)^2 - i[X_1^\ell, X_2^\ell] \right) ,$$

with $X_0 = v\partial_x - \partial_x V(x)\partial_v$, $X_1^\ell = \partial_{v_\ell}$ and $X_2^\ell = iv_\ell/2$. The problem here is that K cannot be viewed as a polynomial of vector fields in some induced representation of a nilpotent Lie algebra. For simplicity, consider the quadratic case $V(x) = x^2/2$ in dimension $n = 1$. We are looking for a graded algebra $\mathcal{G} = \sum_j \mathcal{G}_j$ and for a representation π of the algebra such that:

- $\dim \mathcal{G}_1 = 2$ and there exists a basis Y_1, Y_2 such that $\pi(Y_j) = X_j$.
- $\dim \mathcal{G}_2 = 2$ and $\mathcal{G}_2$ admits as a basis $[Y_1, Y_2]$, Y_0, with $\pi(Y_0) = X_0$.

As a consequence K will appear as

$$K = \pi \left(Y_0 + i[Y_1, Y_2] - Y_1^2 - Y_2^2 \right) .$$

By computing the successive commutators one gets the infinite structure

$\pi(\mathcal{G}_2)$ contains X_0 and i ;
$\pi(\mathcal{G}_3)$ contains ∂_x and ix ;
$\pi(\mathcal{G}_4)$ is reduced to 0 ;
$\pi(\mathcal{G}_5)$ contains iv and ∂_v ;
$\pi(\mathcal{G}_6)$ contains i ;
$\pi(\mathcal{G}_7)$ contains ∂_x and x and so on ...

There are no finite dimensional Lie algebra which permits to do the job like in Folland's example. Hence Helffer-Nourrigat's approach cannot be applied without finding a way to cancel high order irrelevant commutators. This is actually what is done in the analysis of the local hypoellipticity when we neglect the brackets of order $(r + 1)$ but we have here a more difficult global problem. We will see in the next section how to adapt partly the nilpotent techniques.

Second approach:

We can also consider K as an homogeneous operator of order 2 on a stratified nilpotent Lie algebra:

$$K = \sum_\ell i X_2^\ell Y_1^\ell - i Y_2^\ell X_1^\ell - \left((X_1^\ell)^2 + (X_2^\ell)^2 - i[X_1^\ell, X_2^\ell] \right) ,$$

with $X_1^\ell = \partial_{v_\ell}$, $X_2^\ell = i v_\ell/2$, $Y_1^\ell = \partial_{x_\ell}$ and $Y_2^\ell = i(\partial_{x_\ell} V)(x)$. When V is a polynomial of degree m one gets a stratified nilpotent Lie algebra of rank m by taking $X_{1,2}^\ell, Y_{1,2}^\ell \in \mathcal{G}_1$ and K is an homogeneous polynomial of degree 2. (Note that in this case we have $[\mathcal{G}_2, \mathcal{G}_2] = 0$.)

By assuming that the Witten Laplacian is the image $\pi_\tau(P)$ by a representation π_τ of a maximally microlocally hypoellitic operator P on a nilpotent group, there is no difficulty to check that for any representation such that

$$\pi(X_1^\ell) = \partial_{v_\ell} \quad \text{and} \quad \pi(X_2^\ell) = i v_\ell/2 , \tag{9.1}$$

the operator $\pi(K)$ is injective.

The problem comes from the fact that there are non trivial representations for which $\pi(X_1^\ell) = \pi(X_2^\ell) = 0$ and for which therefore $\pi(K) = 0$ is not injective. Hence the nilpotent techniques (see for example [RoSt], [HelNo3], [No1]) cannot be applied directly.

One can think of considering the closed set of representations which satisfy the condition (9.1) but this set is not homogeneous and this leads to difficulties in the adaptation of Helffer-Nourrigat's analysis (see [HelNo3], Chap VIII).

Third approach:
In the previous presentations, one does not recover directly the structure of Witten Laplacian which has to play a role as discussed in Chapter 5 (Indeed it is hidden in the injectivity property for the second approach). We now present an alternative approach which relies as the analysis of the elliptic case on the writing of K with distorted creation-annihilation operators . We consider the $(n + 1) \times (n + 1)$ matrix valued operator $\mathcal{K} = X_0 - X_1^2$ with

$$X_0 = \text{diag}\,(b^*a - a^*b)\,,$$

$$X_1 = i \begin{pmatrix} 0 & b \\ b^* & 0 \end{pmatrix} = i \begin{pmatrix} 0 & \dots & 0 & b_1 \\ \vdots & & \vdots & \vdots \\ 0 & \dots & & b_n \\ b_1^* & \dots & b_n^* & 0 \end{pmatrix}\,,$$

where we used the notations (5.15)(5.16). One recovers the Fokker-Planck operator as the last row of $X_0 - X_1^2$ and a simple calculation gives

$$[X_0, X_1] = X_2 = -i \begin{pmatrix} 0 & a \\ a^* & 0 \end{pmatrix}\,,$$

where the Witten Laplacian in x will appear in the last diagonal element of $-X_2^2$. This has some connection with the differential form structure but it is not clear that this would help for the proof of maximal estimates.

9.2 Maximal Estimates for Some Fokker-Planck Operators

We come back to the analysis of the global estimates for the quadratic model developed in Section 5.5 and show how a nilpotent approach lead to better estimates. The theorem is the following.

Theorem 9.1.
Let us assume that, for $|\alpha| = 2$,

$$|D_x^\alpha V(x)| \le C_\alpha < \nabla V(x) >^{1-\rho_0}\,, \tag{9.2}$$

with

$$\rho_0 > \frac{1}{3}\,, \tag{9.3}$$

and that

$$|\nabla V(x)| \to +\infty\,. \tag{9.4}$$

Then the Fokker-Planck operator K has compact resolvent and we have the inequality

$$||\,|\nabla V(x)|^{\frac{2}{3}}\,u\,||^2 \le C \left(||Ku||^2 + ||u||^2\right)\,, \quad \forall u \in C_0^\infty(\mathbb{R}^n)\,. \tag{9.5}$$

Remark 9.2.
When V is quadratic and non degenerate, we recover all the statements given in Chapter 5 (or at least the results concerning the control at ∞). When $|\nabla V(x)|$ tends to $+\infty$ with a control $< \nabla V(x) > \ge \frac{1}{C} < x >^{\frac{1}{C}}$ and a suitable control of the higher derivatives (see (5.17)), we got already the compactness of the resolvent through the weaker estimate

$$||\,|\nabla V(x)|^{\frac{1}{4}}u\,||^2 \le C \left(||Ku||^2 + ||u||^2\right)\,, \quad \forall u \in C_0^\infty(\mathbb{R}^n)\,. \tag{9.6}$$

But (9.2) is much stronger than (5.17) for the second derivatives of V.

Heuristics.

The approach is the following. One first replaces $\nabla V(x)$ by a constant vector, and prove a global estimate for this model, and then use a partition of unity and control the errors.
Let us present for simplification the approach when $n = 1$ (but this restriction is not important).
The first point (this part is common to the approach given for the quadratic model) is consequently to analyze (with $\nabla V(x_0)$ replaced by w):

$$F_w = (v\partial_x - w\partial_v) + (-\partial_v^2 + \frac{1}{4}v^2 - \frac{1}{2}) \, . \tag{9.7}$$

By partial Fourier transform, we get

$$F_{w,\xi} = (iv\xi - w\partial_v) + (-\partial_v^2 + \frac{1}{4}v^2 - \frac{1}{2}) \, . \tag{9.8}$$

We would like to have uniform estimates with respect to the parameters v, ξ. There is some invariance by rotation in these two variables, so the main parameter is only

$$\lambda = \sqrt{w^2 + \xi^2} \, . \tag{9.9}$$

This leads after a change of variable to the model (see (5.52)):

$$\hat{F}_\lambda = i\lambda t + (-\partial_t^2 + \frac{1}{4}t^2 - \frac{1}{2}) \, , \tag{9.10}$$

for which one can show the existence of a constant $C > 0$, such that for any $\lambda \geq 0$, the following maximal estimate:

$$||u||^2 + ||\hat{F}_\lambda u||^2 \geq \frac{1}{C} \left(\lambda^{\frac{4}{3}} ||u||^2 + \sum_{k+\ell \leq 2} ||D_t^k t^\ell u||^2 \right) \, , \quad \forall u \in \mathcal{S}(\mathbb{R}) \, , \tag{9.11}$$

is satisfied.

Proof of the maximal estimate.
We propose to get it as a consequence of the nilpotent techniques. We will explicitly realize the lifting (this is a particular case of Folland's construction). We introduce the following algebra generated by

$$X_1 = \partial_t \, , \ X_2 = \partial_x - t\partial_y \, , \ X_0 = \partial_z - t\partial_s \, .$$

It defines a rank 3 algebra, whose underlying vector space is $\mathbb{R}^5$, where $\mathcal{G}_1$ is spanned by X_1, X_2, $\mathcal{G}_2$ is spanned by X_0 and $[X_2, X_1] = \partial_y$, and $\mathcal{G}_3$ is spanned by $[X_0, X_1] = \partial_s$.
The operator

$$L = X_1^2 + X_2^2 + X_0$$

is a type 2 Hörmander operator and Rothschild-Stein have shown that this operator is maximally hypoelliptic (see 7.8). So we have in particular

$$||X_0u||^2 + ||X_1^2u||^2 + ||X_1X_2u||^2 + ||X_2^2u||^2 \leq C \, ||Lu||^2 \, , \, \forall u \in C_0^\infty(\mathbb{R}^5) \, ,$$
(9.12)

where the $|| \cdot ||$ denote $L^2(\mathbb{R}^5)$ norms. We emphasize here that it is a global estimate (i.e. better than (8.1)), which can also be deduced from [HelNo2], through the Rockland's criterion. Now one obtains immediately that

$$||\partial_s^2 u|| = ||[X_1, X_0]^2u|| \leq C||u||_{H_G^6}^{hom} \, , \, \forall u \in C_0^\infty(\mathbb{R}^5) \, .$$
(9.13)

Here the $H_G^{2\ell}$, for $\ell \in \mathbb{N}$, are the Sobolev space attached to the vector fields X_1, X_2, X_0 considering X_0 as homogeneous of order 2. Because everything is homogeneous by dilation, we consider the homogeneous semi-norms:

$$C_0^\infty(\mathbb{R}^5) \ni u \mapsto ||u||^{hom,2\ell} = \sum_{|\alpha|_{1,2}=2\ell} ||X^\alpha u||^2 \, ,$$
(9.14)

where $X^\alpha = X_{\alpha_1}X_{\alpha_2}...X_{\alpha_p}$ with $\alpha_i \in \{0,1,2\}$, and

$$|\alpha|_{1,2} = \sum_{j=1}^p \epsilon_j|\alpha_j| \, ,$$

with $\epsilon_j = 1$ if $\alpha_j = 1, 2$ and $\epsilon_j = 2$ if $\alpha_j = 0$.
By complex interpolation theory, we get

$$||(-\partial_s^2)^{\frac{1}{3}}u|| \leq C||u||_{H_G^2}^{hom} \leq C'||Lu|| \, , \, \forall u \in C_0^\infty(\mathbb{R}^5) \, .$$
(9.15)

Using partial Fourier transform and implementing particular choices of families of test functions in inequalities (9.12) and (9.15), we get the inequality

$$||(\hat{F}_\lambda + \frac{1}{2})u||^2 \geq \frac{1}{C}\left(|\lambda|^{\frac{4}{3}}||u||^2 + \sum_{k+\ell=2}||D_t^k t^\ell u||^2\right) \, .$$
(9.16)

This leads to (9.11).

Proof of the theorem.
Coming back to the initial coordinates, we get:

$$||f||_{L^2(\mathbb{R})}^2 + ||F_{w,\xi}f||_{L^2(\mathbb{R})}^2 \geq \frac{1}{C}\left(w^{\frac{4}{3}}||f||_{L^2(\mathbb{R})}^2 + \sum_{k+\ell\leq 2}||D_v^k v^\ell f||_{L^2(\mathbb{R})}^2\right) \, ,$$
(9.17)

for all $f \in C_0^\infty(\mathbb{R})$. Note that by interpolation, this implies also

$$||f||_{L^2(\mathbb{R})}^2 + ||F_{w,\xi}f||_{L^2(\mathbb{R})}^2 \geq \frac{1}{C}\left(w^{\frac{2}{3}}||vf||_{L^2(\mathbb{R})}^2 + |w|^{\frac{2}{3}}||\partial_v f||_{L^2(\mathbb{R})}^2\right) \, , \, \forall f \in C_0^\infty(\mathbb{R}) \, .$$
(9.18)

One can now introduce a partition of unity ϕ_j in the x variable corresponding to a covering by intervals $I(x_j, r(x_j)) =]x_j - r(x_j), x_j + r(x_j)[$ (with the property of uniform finite intersection[1]), where $r(x)$ has the expression

$$r(x) := \delta_0 < \nabla V(x) >^{-\frac{1}{3}}, \qquad (9.19)$$

and where $\delta_0 \geq 1$ is an extra parameter which will be chosen later. So the support of each ϕ_j is contained in $I(x_j, r(x_j))$ and we have

$$\sum_j \phi_j^2(x) = 1. \qquad (9.20)$$

Moreover there exists a constant C, s. t. for all j and for all $\delta_0 \geq 1$,

$$|\nabla \phi_j(x)| \leq \frac{C}{\delta_0} < \nabla V(x_j) >^{\frac{1}{3}}. \qquad (9.21)$$

More precisely, there exists another partition of unity χ_j with support in $I(x_j, 2r(x_j))$, such that

$$\sum \chi_j^2 \leq C,$$

and

$$|\nabla \phi_j(x)| \leq \frac{C}{\delta_0} < \nabla V(x_j) >^{\frac{1}{3}} \chi_j(x). \qquad (9.22)$$

Note that one can find C_{δ_0} such that for $|x_j| \geq C_{\delta_0}$, we have

$$\frac{1}{2} \leq |\nabla V(x)| / |\nabla V(x_j)| \leq \frac{3}{2}, \forall x \in I(x_j, 2r(x_j)).$$

This introduces two types of errors, one is due to the comparison in the interval of $\partial_x V(x)$ and of $\partial_x V(x_0)$ leading to an error in $V''(x_0)r(x_0)$ which has to be small in comparison with the gain $|\partial_x V(x_0)|^{\frac{1}{3}}$. This leads to the assumption that $|V''(x_0)|r(x_0)$ should be controlled by $< \nabla V(x_0) >^{\frac{2}{3}}$. The second type error is due to the partition of unity χ_j. The typical error to control is here $||\phi_j'(x)yf||^2$ which will be controlled, if $r(x)^{-1}$ is controlled by $< \nabla V(x) >^{\frac{1}{3}}$. So this explains our choice of radius and our assumptions on $V''(x)$.

Detailed proof.
Let us now give the detailed proof. We start, for $u \in C_0^\infty(\mathbb{R}^2)$, from

$$\begin{aligned}
||Ku||^2 &= \sum_j ||\phi_j Ku||^2 \\
&= \sum ||K\phi_j u||^2 - \sum ||[K, \phi_j]u||^2 \\
&= \sum ||K\phi_j u||^2 - \sum ||(X_0 \phi_j)u||^2 \\
&= \sum ||K\phi_j u||^2 - \sum ||(\nabla \phi_j)(x)vu||^2 \\
&\geq \sum ||K\phi_j u||^2 - \frac{C^2}{\delta_0^2} \sum ||\chi_j < \nabla V(x_j) >^{\frac{1}{3}} vu||^2.
\end{aligned}$$

---

[1] That is there exists an integer n_0 such that any point is covered by less than n_0 intervals, belonging to the covering family.

Let us now write on the support of ϕ_j:

$$K = K - F_{w_j} + F_{w_j} \, ,$$

with

$$w_j = \nabla V(x_j) \, .$$

We verify that

$$\|(K - F_{w_j})\phi_j u\|^2 = \|\phi_j(x)(\nabla V(x) - \nabla V(x_j))\,\partial_v u\|^2$$
$$\leq C\delta_0^2 < \nabla V(x_j) >^{2-2\rho_0-\frac{2}{3}} \|\phi_j \partial_v u\|^2 \, .$$

These errors have to be controlled by the main term.

We note that we have

$$\|F_{w_j}\phi_j u\|^2 \geq \frac{1}{C}\||\nabla V(x)|^{\frac{2}{3}}\phi_j u\|^2 + \frac{1}{C}\||\nabla V(x)|^{\frac{2}{3}}\phi_j\partial_v u\|^2 + \frac{1}{C}\||\nabla V(x)|^{\frac{2}{3}}\phi_j v u\|^2 \, .$$

Finally, we observe that

$$\|K\phi_j u\|^2 \geq \frac{1}{2}\|F_{w_j}\phi_j u\|^2 - \|(K - F_{w_j})\phi_j u\|^2 \, .$$

Summing up over j, we have obtained the existence of a constant C such that, for any $\delta_0 \geq 1$ and for all $u \in C_0^\infty(\mathbb{R}^2)$,

$$\|u\|^2 + \|Ku\|^2$$
$$\geq \frac{1}{C}\||\nabla V(x)|^{\frac{2}{3}}u\|^2 + \frac{1}{C}\||\nabla V(x)|^{\frac{1}{3}}\partial_v u\|^2 + \frac{1}{C}\||\nabla V(x)|^{\frac{1}{3}}vu\|^2$$
$$-C\delta_0^2\||\nabla V(x)|^{\frac{2}{3}-\rho_0}\partial_v u\|^2 - C\frac{1}{\delta_0^2}\||\nabla V(x)|^{\frac{1}{3}}\partial_v u\|^2$$
$$-C_{\delta_0}(\|vu\|^2 + \|u\|^2 + \|\partial_v u\|^2) \, .$$

If we choose δ_0 large enough, we can achieve the proof by observing that $|\nabla V(x)|$ tends to $+\infty$ and that $\rho_0 > \frac{1}{3}$. ∎

Maximal Microhypoellipticity for Systems and Applications to Witten Laplacians

10.1 Introduction

Although the previous theory for operators on nilpotent groups applies more generally to systems, the criteria which appear for specific first order systems are much more explicit and permit to go much further in the analysis. We develop in more detail a version of the previous point of view, which is adapted to some systems and will show how it leads to new results. Although Kirillov's theory is everywhere behind the formulation of the statements, the language of the nilpotent groups has been in some sense eliminated from the presentation, and the reader of this chapter can survive without any knowledge of this theory.

Here we are mainly inspired by the presentation given by J. Nourrigat in [No1] of results of Helffer-Nourrigat and Nourrigat which appear in a less explicit form in the book [HelNo3]. This is a particular aspect of the large program developed by J. Nourrigat in continuation of [HelNo3] at the end of the 80's for understanding the maximal hypoellipticity of differential systems of order 1 in connection with the chatracterization of subelliptic systems [No1]-[No7]. As we shall see, an interest of this analysis of the maximal hypoellipticity by an approach based on the nilpotent Lie group techniques is that it provides, on the way, global, local or microlocal estimates, leading to sufficient conditions for the compactness of the resolvent or to semiclassical local lower bounds. The comparison with other hypoellipticity results by Maire and their implication to semi-classical analysis will be discussed here and in Section 11.5.

More precisely, our aim is to analyze the maximal hypoellipticity of the system of n first order complex vector fields

$$L_j = (X_j - iY_j), \text{ where } X_j = \partial_{x_j} \text{ and } Y_j = (\partial_{x_j}\Phi(x))\,\partial_t\,, \qquad (10.1)$$

in a neighborhood $\mathcal{V}(0) \times \mathbb{R}_t$ of $0 \in \mathbb{R}^{n+1}$, where $\Phi \in C^\infty(\mathcal{V}(0))$. We will show at the same time how the techniques used for this analysis will lead to some information on the question concerning the Witten Laplacian associated to Φ.

We assume that the real function Φ is such that the rank r Hörmander condition is satisfied for the vector fields $(X_j), (Y_j)$ at $(0,0)$ (and hence at any point $(0, t_0)$ due to the invariance by translation in the t-variable). This is an immediate consequence of the condition:

$$\sum_{1 \leq |\alpha| \leq r} |\partial_x^\alpha \Phi(0)| > 0 . \tag{10.2}$$

Let us start by extending the previous notion of maximal hypoellipticity to systems. By maximal hypoellipticity for the system (10.1), we mean the existence of the inequality:

$$\sum_j ||X_j u||^2 + \sum_j ||Y_j u||^2 \leq C \left(\sum_j ||L_j u||^2 + ||u||^2 \right) , \forall u \in C_0^\infty(\mathcal{V}(0) \times \mathbb{R}_t) . \tag{10.3}$$

The symbol of the system is the map:

$$\begin{aligned} T^*(\mathcal{V}(0) \times \mathbb{R}) \setminus \{0\} &\ni (x, t, \xi, \tau) \\ &\mapsto \sigma(L)(x, t, \xi, \tau) := \left(i\xi_j + \tau(\partial_{x_j} \Phi)(x) \right)_{j=1,\dots,n} \in \mathbb{C}^n . \end{aligned} \tag{10.4}$$

Note that it is also the the principal symbol, which is homogeneous of degree 1:

$$\sigma(L)(x, t, \rho\xi, \rho\tau) = \rho\, \sigma(x, t, \xi, \tau) , \forall \rho > 0 . \tag{10.5}$$

The characteristic set is then by definition the set of zeroes of (the principal symbol of) $\sigma(L)$:

$$\sigma(L)^{-1}(0) = \{(x, t, \xi, \tau) \in T^*(\mathbb{R}^{n+1}) \setminus \{0\} \mid \xi = 0 , \nabla\Phi(x) = 0\} , \tag{10.6}$$

and is consequently a conic subset of $T^*(\mathcal{V}(0) \times \mathbb{R}) \setminus \{0\}$. Outside this set the system is microlocally elliptic (its (principal) symbol does not vanish) and hence (maximally) microlocally hypoelliptic. So the local (maximal) hypoellipticity will result of the microlocal analysis in the neighborhood of the characteristic set, which has actually two connected components defined by $\{\pm\tau > 0\}$. So we are more precisely interested in the microlocal hypoellipticity in a conic neighborhood $V_\pm$ of $(x, t; \xi, \tau) = (0; 0, \pm 1)$, that is with the microlocalized version of the inequality (10.3), which writes:

$$\begin{aligned} \sum_j ||\chi_\pm(x, t, D_x, D_t) X_j u||^2 + \sum_j ||\chi_\pm(x, t, D_x, D_t) Y_j u||^2 \\ \leq C \left(\sum_j ||L_j u||^2 + ||u||^2 \right) , \forall u \in C_0^\infty(\mathcal{V}(0) \times \mathbb{R}_t) , \end{aligned} \tag{10.7}$$

where $\chi_\pm$ is a classical pseudo-differential operator of order 0 which localizes in $V_\pm$ (whose principal symbol should be elliptic at $(0; 0, \pm 1)$).

Actually, due to the invariance of the problem with respect to the t variable, we will more precisely look for the existence of an inequality which is local in x but global in the t variable:

$$\sum_j ||\chi_\pm(x, D_x, D_t) X_j u||^2 + \sum_j ||\chi_\pm(x, D_x, D_t) Y_j u||^2$$
$$\leq C \left(\sum_j ||L_j u||^2 + ||u||^2 \right), \ \forall u \in C_0^\infty(\mathcal{V}(0) \times \mathbb{R}_t). \tag{10.8}$$

This will permit to consider the partial Fourier transform with respect to t in order to analyze the problem.

It is interesting to observe that in the right hand side of (10.8), we can write:

$$\sum_j ||L_j u||^2 = \sum_j \langle L_j^* L_j u \mid u \rangle, \tag{10.9}$$

and that

$$\sum_j L_j^* L_j = - \left(\sum_j X_j^2 + \sum_j Y_j^2 - i \sum_j [X_j, Y_j] \right). \tag{10.10}$$

10.2 Microlocal Hypoellipticity and Semi-classical Analysis

In this section, we show how global estimates for operators with t-independent coefficients lead after partial Fourier transform with respect to the t-variable to semiclassical results.

10.2.1 Analysis of the Links

Observing the translation invariance with respect to t of the system (10.1), it is natural to ask for the existence of $C > 0$, such that, for any $\tau \in \mathbb{R}$, the inequality:

$$\sum_j ||\pi_\tau(X_j) v||^2 + \sum_j ||\pi_\tau(Y_j) v||^2 \leq C \left(\sum_j ||\pi_\tau(L_j) v||^2 + ||v||^2 \right), \tag{10.11}$$

is satisfied for all $v \in C_0^\infty(\mathcal{V}(0))$, where $\mathcal{V}(0)$ is a neighborhood of 0 in $\mathbb{R}^n$ and

$$\pi_\tau(L_j) = \pi_\tau(X_j) - i\pi_\tau(Y_j) = \partial_{x_j} + \tau(\partial_j \Phi)(x). \tag{10.12}$$

The emphasize that the constant C above is independent of τ.

Using the partial Fourier transform with respect to t, one can indeed show that the proof of (10.11) uniformly with respect to τ is the main point for getting the maximal estimate. We now give two remarks:

1. The estimate (10.11) is trivial for τ in a bounded set.
2. Depending on which connected component of the characteristic set is concerned, we have to consider the inequality for $\pm\tau \geq 0$ (τ large).

From now on, we choose the $+$ component and assume

$$\tau > 0 \tag{10.13}$$

for simplicity. In any case, changing Φ into $-\Phi$ exchanges the roles of $\tau > 0$ and $\tau < 0$, so there is no loss of generality in this choice. If we introduce the semi-classical parameter by:

$$h = \frac{1}{\tau} , \tag{10.14}$$

the inequality (10.11) becomes, after division by τ^2:

$$\sum_j ||(h\partial_{x_j})v||^2 + \sum_j ||(\partial_{x_j}\Phi)v||^2 \leq C \left(\langle \Delta_{\Phi,h}^{(0)} v \mid v \rangle + h^2||v||^2 \right) , \tag{10.15}$$

for all $v \in C_0^\infty(\mathcal{V}(0))$, where

$$\Delta_{\Phi,h}^{(0)} = -h^2\Delta + |\nabla\Phi|^2 - h\Delta\Phi , \tag{10.16}$$

is the semi-classical Witten Laplacian on functions introduced in (1.2).

Hörmander's condition gives as a consequence of the microlocal subelliptic estimate (cf also [BoCaNo]) the existence of $\mathcal{V}(0)$, $h_0 > 0$ and $C > 0$ such that:

$$h^{2-\frac{2}{r}}||v||^2 \leq C \left(\sum_j ||(h\partial_{x_j})v||^2 + \sum_j ||(\partial_{x_j}\Phi)v||^2 \right) \tag{10.17}$$

for $h \in]0, h_0]$ and $v \in C_0^\infty(\mathcal{V}(0))$.

So we finally obtain the existence of $\mathcal{V}(0)$, $h_0 > 0$ and $C > 0$ such that:

$$h^{2-\frac{2}{r}}||v||^2 \leq C \langle \Delta_{\Phi,h}^{(0)} v \mid v \rangle , \ \forall v \in C_0^\infty(\mathcal{V}(0)) , \tag{10.18}$$

for $h \in]0, h_0]$. So the maximal microhypoellipticity (actually the subellipticity would have been enough) in the "+" component implies some semi-classical localized lower bound for the semi-classical Witten Laplacian of order 0.

Remark 10.1.
We refer to [HelNo4] for an analysis of a connected semi-classical subelliptic uncertainty principle.

Of course, many semi-classical results can be obtained by other techniques, particularly in the case when Φ is a Morse function with a critical point at 0. This will be analyzed in great detail in Section 14.4. So it is more in degenerate cases that these "old" microlocal results can be relooked for giving "new" semi-classical results. In the case when $r = 2$, we will see in particular that, when Φ is a Morse function, then the condition for the maximal microlocal hypoellipticity at $(0; 0, 1)$ is that Φ is not a local minimum. According to Maire's results in [Mai1] a similar condition occurs more generally when Φ is analytic.

For the discussion of the different approaches, it is convenient to introduce the

Definition 10.2.
The semiclassical Witten Laplacian $\Delta^{(0)}_{\Phi,h}$ is said δ-subelliptic, $0 \leq \delta < 1$, in an open set Ω, if[1] there exist $C > 0$ and $h_0 > 0$ such that the estimate,

$$h^{2\delta} \|v\|^2 \leq C \|d^{(0)}_{\Phi,h} v\|^2 , \tag{10.19}$$

holds uniformly for all $h \in (0, h_0]$ and $v \in \mathcal{C}^\infty_0(\Omega)$.

The estimate (10.18) says that $\Delta^{(0)}_{\Phi,h}$ is $(1 - \frac{1}{r})$-subelliptic in a neighborhood of $x = 0$. If one goes back to the system with $\tau = \frac{1}{h}$, the δ-subellipticity of $\Delta^{(0)}_{\Phi,h}$ gives for the system L_j:

$$\tau^{2-2\delta} \|v\|^2 \leq C \left(\sum_j \|\pi_\tau(L_j)v\|^2 \right) , \quad \forall \tau > 0 ,$$

which means that the system (L_j) is microlocally hypoelliptic near $(0; 0, +1)$ with loss of δ derivatives, in comparison with the elliptic case where $\delta = 0$.

10.2.2 Analysis of the Microhypoellipticity for Systems

Let us now express what the group theoretical criteria will give for the semi-classical Witten Laplacian.

Definition 10.3.
We denote by $\mathcal{L}_0$ the set of all polynomials P of degree less or equal to r vanishing at 0 ($P \in E_r$) such that there exists a sequence $x_n \to 0$, $\tau_n \to +\infty$ and $d_n \to 0$ such that:

$$d_n^{|\alpha|} \tau_n (\partial^\alpha_x \Phi)(x_n) \to \partial^\alpha_x P(0) . \tag{10.20}$$

Remark 10.4.
If the Hörmander condition of rank r can not be improved at 0 (or equivalently, choose the smallest r such that (10.2) is satisfied), we have:

$$\lim_{n \to +\infty} d_n^r \tau_n \neq 0 . \tag{10.21}$$

---

[1] Strictly speaking, one should have added: for any compact K in Ω, there exists $C_K...$, but this is unimportant here because Ω is not given a priori but will be some neighborhood of a given point.

In the case, when Φ is a Morse function, the set $\mathcal{L}_0$ is simply the quadratic approximation of Φ at 0 up to a multiplicative positive constant. Now, the translation of Helffer-Nourrigat's Theorem [HelNo3] provides the semiclassical estimate:

Theorem 10.5.
We assume that (10.2) is satisfied at rank r. Then, if the condition:
No polynomial in $\mathcal{L}_0$ except 0 has a local minimum at the origin,
then the operator $\Delta_{\Phi,h}^{(0)}$ is $(1 - \frac{1}{r})$-subelliptic in a neighborhood of $x = 0$ (inequality (10.18) holds for h small enough). Moreover, the condition is necessary for getting the maximal estimate (10.15).

Remark 10.6.
There is an equivalent way to express the condition in Theorem 10.5. There exists a neighborhood V of 0 and two constants d_0 and c_0, such that:

$$\inf_{|x-x_1|\leq d}(\Phi(x) - \Phi(x_1)) \leq -c_0 \sup_{|x-x_1|\leq d} |\Phi(x) - \Phi(x_1)| \,,$$

for all $x_1 \in V$ and for all $d \in [0, d_0[$. We refer also to F. Trèves [Tr2] (and [No1]-[No7], [HelNo3] and [Mai1]).

Remark 10.7.
When Φ is a Morse function, and if Φ has a local minimum at a point x_{min}, the implementation in (10.18) of the trial function $\chi \exp -\frac{\Phi}{h}$, where χ is a cut-off function localizing in the neighborhood of x_{min}, shows that there are no hope to have a subelliptic estimate . The right hand side in (10.18) becomes indeed exponentially small $\mathcal{O}(\exp -\frac{\alpha}{h})$, for some $\alpha > 0$. This argument works more generally under the weaker assumption that Φ has isolated critical points, without assuming the Morse property.

In connection with previous work by F. Trèves [Tr2], Maire's results [Mai1] suggest the

Conjecture 10.8.
Under the assumption that Φ is analytic and that Φ has no local minimum at the origin, then $\Delta_{\Phi,h}^{(0)}$ is δ-subelliptic in a neighborhood of 0 for some $\delta \in [0, 1)$.

Maire's result on hypoellipticity relies crucially on the Lojaciewicz's inequality, saying that for an analytic function Φ defined in say a neighborhood of 0, then there exists a constant $C > 0$ and $\theta \in]0, 1[$ such that in a possibly smaller neighborhood $\mathcal{V}_0$ of 0, we have:

$$|\nabla\Phi(x)| \geq C \, |\Phi(x)|^\theta \,, \quad \forall x \in \mathcal{V}_0 \,.$$

Actually the work of Maire [Mai1] is concerned with more general systems for which τ lies in a multidimensional space. The situation met here is one dimensional, but the proof of Maire leads[2] only to a weak form of subellipticity, implying microlocal hypoellipticity but giving only an L^∞ version of (10.19). Note that Maire has also shown that this "subellipticity" is not equivalent to maximal hypoellipticity of the system, when $n > 1$ (cf Proposition 20 in [Mai2]). For example (cf Example 1.2, p. 56, in [Mai4]), the system

$$\begin{cases} L_1 = \partial_{x_1} - i\left((2\ell+1)x_1^{2\ell} - x_2^2\right)\partial_t \,, \\ L_2 = \partial_{x_2} + 2ix_1x_2\partial_t \,, \end{cases} \tag{10.22}$$

is, according to Maire[3] L^2-microlocally subelliptic with loss of $\frac{2\ell}{2\ell+1}$ derivatives. According to Definition 10.2, this implies that the Witten Laplacian $\Delta_{\Phi,h}^{(0)}$, with

$$\Phi = x_1^{2\ell+1} - x_1 x_2^2 \,,$$

is $\frac{2\ell}{2\ell+1}$-subelliptic in a neighborhood of $x = 0$. Actually this exponent $\delta = \frac{2\ell}{2\ell+1}$ can not be improved and we will come back to this point in Section 11.5. Meanwhile the microlocal maximal hypoellipticity would give $\delta = 1 - 1/3 = 2/3$ because the Hörmander condition is satisfied at rank 3. Hence the above system cannot be microlocally maximally hypoelliptic for $\ell > 1$ at $(0;0,1)$ and the semiclassical estimate (10.17) does not hold.

10.3 Around the Proof of Theorem 10.5

The proof is based on a priori estimates obtained by a recursion argument strongly related to Kirillov's theory. All this section is strongly inspired by the presentation of J. Nourrigat [No1]. We first observe that the set $\mathcal{L}_0$ introduced in Definition 10.3 has some stability[4] properties.

Proposition 10.9.
The set $\mathcal{L}_0$ has the following properties:

1. If $P \in \mathcal{L}_0$ and $y \in \mathbb{R}^n$, then the polynomial defined by

$$Q(x) = P(x+y) - P(y), \forall x \in \mathbb{R}^n \,,$$

is also in $\mathcal{L}_0$.
2. If $P \in \mathcal{L}_0$ and $\lambda > 0$, then $Q(x) = P(\lambda x)$ is also in $\mathcal{L}_0$.
3. $\mathcal{L}_0$ is a closed subset of E_r.

Definition 10.10.
A set in E_r satisfying the three conditions of Proposition 10.9 will be called canonical set.

---

[2] We missed this point in the preliminary version of these notes and thank H.M. Maire and M. Derridj for pointing out this difficulty.
[3] Personal communication ([Mai2] provides a weaker result).
[4] which are actually the translation of the group theoretical properties appearing before the definition of Γ_{x_0,ξ_0} in Conjecture 8.3,

To each polynomial $P \in E_r$, we can associate a system of differential operators in $\mathbb{R}^n$ by

$$\pi_P(X_j) = D_{x_j} \, , \ \pi_P(Y_j) = \partial_{x_j} P \, , \ \pi_P(L_j) = \pi_P(X_j) - i\pi_P(Y_j) \, . \quad (10.23)$$

Reduction of the number of variables

In order to prove maximal estimates like:

$$\sum_j ||D_{x_j} u||^2 + \lambda^2 ||(\partial_{x_j} P) \, u||^2 \leq C \sum_j ||(D_{x_j} - i\lambda(\partial_{x_j} P)) \, u||^2 \, , \quad (10.24)$$

we should also consider operators obtained from above by reduction of the number of variables. After a suitable linear change of variables $x \mapsto t$, we get an integer $k = k(P)$ and a family depending on $\tau \in \mathbb{R}^{n-k}$ of operators on $\mathbb{R}^k$

$$\pi_{\hat{P},\tau}(X_j) = D_{t_j} \, , \ \text{for } j = 1, \ldots, k \, , \ \ \pi_{\hat{P},\tau}(X_j) = \tau_j \, , \ \text{for } j = k+1 \ldots n \, ,$$
$$\pi_{\hat{P},\tau}(Y_j) = \partial_{t_j} \hat{P} \, , \ \text{for } j = 1, \ldots, n \, ,$$

$$(10.25)$$

with $\hat{P}(t) = P(x)$. Note that $(\partial_{t_j} \hat{P})$ is a polynomial which is independent of the variables $t_{k+1}, \cdots, t_n$ for $j = 1, \ldots, n$. When $k(P) = n$, π_P corresponds to an irreducible representation. When $k(P) < n$, what we have briefly described is how one can decompose π_P as an Hilbertian integral (in the τ-variables) of irreducible representations $\pi_{\hat{P},\tau}$.

It is not to difficult to show:

Proposition 10.11.
If a polynomial $P \in E_r$, $P \neq 0$ has no local minimum in $\mathbb{R}^n$, then, for any $\tau \in \mathbb{R}^{n-k}$, the system $\pi_{\hat{P},\tau}(L_j)$ is injective[5] on $\mathcal{S}(\mathbb{R}^{k(P)})$.

One observes indeed that for $\tau = 0$ (which is the unique non trivial case), a solution of $\pi_{\hat{P},\tau}(L_j)u = 0$ is necessarily (up to a multiplicative constant) $u = \exp -\hat{P}(t)$. The function u being in $\mathcal{S}(\mathbb{R}^{k(P)})$ and positive should have a maximum in contradiction with the property of $\hat{P}$.

The next proposition appears in a different but equivalent language in [HelNo3] and is the core of the proof:

Proposition 10.12.
Let $\mathcal{L}$ be a canonical subset of E_r. We assume that for any $P \in \mathcal{L} \setminus \{0\}$ and for any $\tau \in \mathbb{R}^{n-k(P)}$ the system $\pi_{\hat{P},\tau}(L_j)$ is injective. Then there exists a constant $c_0 > 0$ such that:

$$\sum_j ||\pi_P(X_j)u||^2 + \sum_j ||\pi_P(Y_j)u||^2 \leq c_0 \sum_j ||\pi_P(L_j)u||^2 \, , \quad (10.26)$$

for all $P \in \mathcal{L}$ and for all $u \in \mathcal{S}(\mathbb{R}^n)$.

---

[5] In other words, one can not find any trivial $u \in \mathcal{S}(\mathbb{R}^{k(P)})$ such that $\pi_{\hat{P},\tau}(L_j)u = 0$ for $j = 1, \ldots, n$.

We shall not give the complete, actually rather involved, proof of the proposition but we would like to emphasize various points of the proof which will actually have also consequences for the analysis of the Witten Laplacian.

1. Continuity.

An intermediate step is to show that

$$\sum_j ||\pi_{\hat{P},\tau}(X_j)u||^2 + \sum_j ||\pi_{\hat{P},\tau}(Y_j)u||^2 \leq c_0(\hat{P},\tau) \sum_j ||\pi_{\hat{P},\tau}(L_j)u||^2 \;, \quad (10.27)$$

for all $u \in \mathcal{S}(\mathbb{R}^{k(P)})$, and to control the uniformity of the constant $c_0(\hat{P},\tau)$ with respect to τ. This leads to (10.26) for some constant $c_0(P)$ depending on $P \in \mathcal{L} \setminus \{0\}$. Then the second **difficult** point is to control the uniformity of the constant $c_0(P)$ with respect to P.

2. Control at ∞.

We would like to first mention the following key lemma:

Lemma 10.13.
With

$$R_P(x) = \sum_{1 \leq |\alpha| \leq r} |\partial_x^\alpha P(x)|^{\frac{1}{|\alpha|}} \;,$$

there exists a constant $c_1 > 0$ such that

$$||R_P u||^2 \leq c_1 \left(\sum_j ||\pi_P(X_j)u||^2 + ||\pi_P(Y_j)u||^2 \right) \;, \quad (10.28)$$

for all $P \in E_r$ and for all $u \in \mathcal{S}(\mathbb{R}^n)$.

Remark 10.14.
This last estimate is much more accurate than what is obtained by the approach of Helffer-Mohamed [HelMo] presented in Section 3.3.

3. Inequality with remainder.

The proof of this proposition being by induction on the rank. The argument which permit to pass from rank $(r-1)$ to rank r is the following intermediate result.

Proposition 10.15.
If $\mathcal{L}$ is canonical and if (10.26) is valid for all $P \in \mathcal{L} \cap E_{r-1}$, then there exists $c_1 > 0$ such that:

$$\sum_j ||\pi_P(X_j)u||^2 + \sum_j ||\pi_P(Y_j)u||^2 \leq c_1 \left(\sum_j ||\pi_P(L_j)u||^2 + [P]_r^2||u||^2 \right) \;,$$

$$(10.29)$$

for all $P \in \mathcal{L}$ and for all $u \in \mathcal{S}(\mathbb{R}^n)$, with

$$[P]_r = \left[\sum_{|\alpha|=r} \left| P^{(\alpha)}(0) \right| \right]^{1/r}.$$

10.4 Spectral By-products for the Witten Laplacians

10.4.1 Main Statements

We shall use Proposition 10.15 in the following way. For a polynomial $\Phi \in E_r$, we denote by $\mathcal{L}_\Phi$ the smallest canonical closed set containing Φ.

Theorem 10.16.
Let $\Phi \in E_r$ and let us assume that:

1. *The representation π_Φ is irreducible[6].*
2. *The canonical set $\mathcal{L}_\Phi \cap E_{r-1}$ does not contain any non zero polynomial having a local minimum.*

Then the Witten Laplacian $\Delta_\Phi^{(0)}$ has compact resolvent. Moreover we have maximal estimates for the system $d_\Phi^{(0)}$ and for the corresponding Laplacian $\Delta_\Phi^{(0)}$.

1. The condition of irreducibility is necessary for having compact resolvent (see [HelNo3] and the discussion in Chapter 11).
2. Let us observe that the condition is stable if one replace Φ by $\lambda\Phi$, with $\lambda > 0$. It will be the same for the results of Helffer-Nier [HelNi1], that will be discussed in Chapter 11.
3. It seems reasonable that, by extra work, one could treat the case of more general functions Φ, which are no more polynomials. We shall present a more pedestrian approach in Chapter 11 in the case when the function Φ is a sum of homogeneous functions at ∞.
4. Another interesting problem would be to determine in the spirit of Helffer-Mohamed [HelMo] the essential spectrum of the Witten Laplacian, when the condition of compactness is not satisfied. We are indeed mainly interested, for the applications to the Fokker-Planck operator, in determining the existence of a gap between 0 and the lowest non zero eigenvalue.
5. Another idea which could be efficient would be to analyze in the same way the corresponding Witten Laplacian on the one-forms. The existence of a gap for $\Delta_\Phi^{(0)}$ should be, in the semi-classical limit, a consequence of the microhypoellipticity of the operator on one forms:

---

[6] We recall that this condition is equivalent to $k(\Phi) = n$ or to the property that $\sum_{|\alpha|>0} |D_x^\alpha \Phi(x)| \to +\infty$ as $|x| \to +\infty$.

$$\Big(\sum_j L_j^* L_j\Big) \times I - i \operatorname{Hess} \varPhi\, \partial_t \,,$$

with L_j defined in (10.1).

6. Other examples of this type are considered in [GHH].

The above theorem could look rather difficult to apply to concrete examples. This is indeed true but let us now give examples showing that one can recover some of the results already obtained and also new results. Let us analyze how one can apply our Theorem 10.16

10.4.2 Applications for Homogeneous Examples

Let us assume that $\varPhi$ is an homogeneous polynomial of degree r without translational invariance (here the point of view developed in Section 8.3 with the induced representation $\pi_{0,\mathcal{H}}$ is also working). In order to apply Theorem 10.16, we have to determine the set $\mathcal{L}_\varPhi \cap E_{r-1}$. One has consequently to determine the polynomials P_∞ of order $r-1$ appearing as limits:

$$P_\infty = \lim_{n\to+\infty} \left(\lambda_n^r \varPhi(\cdot + h_n) - \lambda_n^r \varPhi(h_n)\right) \,,$$

for some sequence (λ_n, h_n) with

$$\lambda_n \to 0 \,.$$

The coefficients of this limiting polynomial P_∞ should satisfy:

$$\lim_{n\to+\infty} \lambda_n^r (\partial_x^\alpha \varPhi)(h_n) = (\partial_x^\alpha P_\infty)(0) \,.$$

10.4.2.1. Elliptic case.

Let us treat the "elliptic" case corresponding to

$$\nabla \varPhi(x) \neq 0 \,, \ \forall x \neq 0 \,. \tag{10.30}$$

Let us show that the limit polynomial is necessarily of degree 1. If it was not the case, there would be some α with $|\alpha| \geq 2$, such that

$$\lim_{n\to+\infty} |\lambda_n^r (\partial_x^\alpha \varPhi)(h_n)| > 0 \,.$$

This would imply the existence of a constant $C_1 > 0$ such that:

$$\frac{1}{C_1} \leq |h_n|^{r-|\alpha|} |\lambda_n|^r \,,$$

and hence $|h_n| \to +\infty$.

We now use the ellipticity. We have, for suitable positive constants C_1, C_2, and C_3,

$$C_3 \geq |\nabla\Phi(h_n)||\lambda_n|^r \geq \frac{1}{C_2}|h_n|^{r-1}\lambda_n^r \geq \frac{1}{C_1 C_2}|h_n|^{|\alpha|-1}\,.$$

This leads to a contradiction with $|\alpha| > 1$ and $\lim_{n\to+\infty} |h_n| = +\infty$. So P_∞ is a polynomial of degree one, which clearly can not have any local minimum.

This case can also be treated more directly by observing that, under assumption (10.30), $|\nabla\Phi(x)| \to +\infty$ as $|x| \to +\infty$ and by observing that $\Delta\Phi$ is of lower order. We indeed immediately obtain that:

$$\lim_{|x|\to+\infty} \left(|\nabla\Phi(x)|^2 - \Delta\Phi(x)\right) = +\infty\,.$$

So at this point, the applications are reconforting but rather poor.

10.4.2.2. Generic non-elliptic case

We assume now that

$$|\nabla\Phi|^{-1}(0) \cap (\mathbb{R}^n \setminus \{0\}) \neq \emptyset\,, \tag{10.31}$$

and introduce the non degeneracy condition:

$$\sum_{1\leq|\alpha|\leq 2} |\Phi^{(\alpha)}(x)| \neq 0\,, \ \forall x \neq 0\,. \tag{10.32}$$

It is easy to see that, under this condition, all the limiting polynomials in $\mathcal{L}_\Phi \cap E_{r-1}$ should be of order less than 2. Because Φ is homogeneous, we have the additional following property:

$$\forall\omega \in (\nabla\Phi)^{-1}(\{0\}) \cap \mathbb{S}^{n-1}\,, \ \Phi''(\omega)\cdot\omega = 0\,. \tag{10.33}$$

Then we have the following

Proposition 10.17.
Under assumptions (10.32) and if, for all $\omega \in (\nabla\Phi)^{-1}(\{0\}) \cap \mathbb{S}^{n-1}$, the Hessian $\Phi''(\omega)$ restricted to $(\mathbb{R}\omega)^\perp$ is non degenerate and not of index 0, then the corresponding Witten Laplacian $\Delta_\Phi^{(0)}$ has a compact resolvent.

Proof.
First we observe that the non degeneracy condition implies that all the limiting polynomials are of degree ≤ 2. Their homogeneous part is up to some positive multiplicative factor the Hessian of Φ at ω, for some ω such that $\nabla\Phi(\omega) = 0$. The assumption says that the signature of this Hessian can not be $(0, +, \ldots, +)$. But this was the only case when the Hessian can have a local minimum. In the other case, the quadratic polynomial has no local minimum and this property cannot be perturbed by linear terms.

Remark 10.18.
The proposition is still true when a polynomial of lower order is added to Φ. We will see in the next subsection that, in contrary, the compactness of the resolvent for the Witten Laplacian $\Delta_\Phi^{(0)}$ could be true for some non homogeneous Φ whose homogeneous principal part does not satisfy the conditions of Proposition 10.17.

Let us consider as an example the case:

$$\Phi_\varepsilon(x_1, x_2) = \varepsilon x_1^2 x_2^2 , \quad \text{with } \varepsilon = \pm 1 , \tag{10.34}$$

and let us determine more explicitly all the limiting polynomials. Writing a limiting polynomial in the form:

$$P_\infty(x) = \ell_1 x_1 + \ell_2 x_2 + \frac{1}{2}\ell_{11} x_1^2 + \ell_{12} x_1 x_2 + \frac{1}{2}\ell_{22} x_2^2 ,$$

we obtain, using the same notations as in the previous example

$$\begin{aligned}
2\varepsilon \lambda_n^4 x_{1,n} x_{2,n}^2 &\to \ell_1 , \\
2\varepsilon \lambda_n^4 x_{1,n}^2 x_{2,n} &\to \ell_2 , \\
2\varepsilon \lambda_n^4 x_{2,n}^2 &\to \ell_{11} , \\
4\varepsilon \lambda_n^4 x_{1,n} x_{2,n} &\to \ell_{12} , \\
2\varepsilon \lambda_n^4 x_{1,n}^2 &\to \ell_{22} .
\end{aligned} \tag{10.35}$$

The limits should obviously verify:

$$\begin{aligned}
\ell_{12}^2 &= 4\ell_{11}\ell_{22} , \\
\ell_1^2 \ell_{22} &= \ell_2^2 \ell_{11} .
\end{aligned} \tag{10.36}$$

The only non trivial polynomials which can have a local minimum should be effectively of order 2. So we assume that

$$\sum_{i,j} \ell_{ij}^2 \neq 0 . \tag{10.37}$$

This is only possible, due to the first line of (10.36), if

$$\ell_{11}^2 + \ell_{22}^2 \neq 0 . \tag{10.38}$$

Because $\lambda_n \to 0$, we deduce from this property that

$$x_{1,n}^2 + x_{2,n}^2 \to +\infty . \tag{10.39}$$

We now show that

$$\ell_{12} = 0 . \tag{10.40}$$

If it was not the case, we get immediately a contradiction between (10.39), the fourth line of (10.35) and the first or the second line of (10.35). Using the first line of (10.36) we get that

$$\ell_{11}\ell_{22} = 0 . \tag{10.41}$$

Using finally the second line of (10.37) and (10.38), we have finally found two different cases:

$$\begin{aligned}
P_\infty(x) &= \tfrac{\gamma}{2}x_1^2 + \ell_1 x_1 , \quad \text{(with } \varepsilon\gamma > 0) , \\
P_\infty(x) &= \tfrac{\gamma}{2}x_2^2 + \ell_2 x_2 , \quad \text{(with } \varepsilon\gamma > 0) .
\end{aligned} \tag{10.42}$$

The conclusion is

Proposition 10.19.
For $\Phi(x_1, x_2) = -x_1^2 x_2^2$, the corresponding Witten Laplacian has a compact resolvent. Moreover, we have the following maximal estimates:

$$||\partial_{x_1} u||^2 + ||\partial_{x_2} u||^2 + ||x_1 x_2 u||^2 \leq C \langle \Delta_\Phi^{(0)} u \mid u \rangle$$

and

$$||\partial_{x_1}^2 u||^2 + ||\partial_{x_1 x_2} u||^2 + ||\partial_{x_2}^2 u||^2 + ||x_1^2 x_2^2 u||^2 \leq C \langle \Delta_\Phi^{(0)} u \mid u \rangle .$$

We recover in this case what we get directly from Proposition 3.1 but note that this new proof gives however a stronger estimate, which does not result from this proposition.

When $\varepsilon = 1$, one can find a non trivial polynomial having a local minimum. The criterion does not apply. We will show later that the operator actually does not have a compact resolvent.

10.4.3 Applications for Non-homogeneous Examples

As an illustration of the second comment in Remark 10.18, let us analyze two examples.

10.4.3.1. The example: $x_1^2 x_2^2 + \epsilon(x_1^2 + x_2^2)$, when $\epsilon \neq 0$.

The proof is a variant of the previous computation. We just mention the differences. (10.35) is replaced by

$$\begin{aligned}
2\lambda_n^4 x_{1,n} x_{2,n}^2 + 2\epsilon\lambda_n^2 x_{1,n} &\to \ell_1 , \\
2\lambda_n^4 x_{1,n}^2 x_{2,n} + 2\epsilon\lambda_n^2 x_{2,n} &\to \ell_2 , \\
2\lambda_n^4 x_{2,n}^2 + 2\epsilon\lambda_n^2 &\to \ell_{11} , \\
4\lambda_n^4 x_{1,n} x_{2,n} &\to \ell_{12} , \\
2\lambda_n^4 x_{1,n}^2 + 2\epsilon\lambda_n^2 &\to \ell_{22} .
\end{aligned} \tag{10.43}$$

One shows again that $\ell_{12} = 0$ and that we have then the following cases to consider:

$$\ell_{11} > 0 , \quad \ell_{22} = 0 , \quad \ell_2 = \pm\epsilon\sqrt{2\ell_{11}} ,$$

and

$$\ell_{22} > 0 \, , \ \ell_{11} = 0 \, , \ \ell_1 = \pm\epsilon\sqrt{2\ell_{22}} \, .$$

So the limiting non-trivial polynomials (up to translation and renormalization) have the form

$$\frac{\gamma}{2}x_1^2 \pm \epsilon\sqrt{2\gamma}x_2 \, ,$$

(with $\gamma > 0$) and

$$\frac{\gamma}{2}x_2^2 \pm \epsilon\sqrt{2\gamma}x_1 \, ,$$

(with $\gamma > 0$), and can not have local minima. So we have obtained

Proposition 10.20.
For $\epsilon \neq 0$, the Witten Laplacian attached to $\Phi = x_1^2 x_2^2 + \epsilon(x_1^2 + x_2^2)$ has a compact resolvent.

We will give another proof of this property for $\epsilon > 0$ (see Theorems 11.3 and 11.10). Note also that in the case when $\Phi = x_1^2 x_2^2 + \epsilon(x_1^2 - x_2^2)$ the Witten Laplacian has also a compact resolvent.

10.4.3.2. The example $\Phi_\varepsilon = (x_1^2 - x_2)^2 + \varepsilon x_2^2$ in $\mathbb{R}^2$ with $\varepsilon \in \mathbb{R}$.

As we will now check, Theorem 10.16 gives a complete answer:

Proposition 10.21.
$\Delta_{\Phi_\varepsilon}^{(0)}$ *has a compact resolvent if and only if $\varepsilon \neq 0$ and $0 \in \sigma_{ess}\left(\Delta_{\Phi_0}^{(0)}\right)$.*

The last statement is obtained by considering a Weyl sequence[7] $(n \geq 2)$ of functions $(x_1, x_2) \mapsto u_n(x_1, x_2) := \chi(\frac{x_1 - n^2}{n}) \exp -\Phi_0(x_1, x_2)$ and using the change of variable

$$y_1 = x_1 \, , \ y_2 = x_2 - x_1^2 \, ,$$

which preserves the Lebesgue measure.

In order to check the hypotheses of Theorem 10.16, we have to specify the elements of $\mathcal{L}_{\Phi_\varepsilon} \cap E_3$, that is to consider the polynomials of degree ≤ 3 which are limits of

$$Q_{\lambda,y}(x) = \Phi_\varepsilon(\lambda(x + y)) - \Phi_\varepsilon(\lambda y) \, .$$

It suffices to look at the possible limits of the derivatives at 0, $\partial_x^\alpha[\Phi_\varepsilon(\lambda x)]$. We get:

$\alpha = (4, 0)$: $24\lambda^4 \to 0$. We are looking at polynomials with degree ≤ 3 which forces the limit to be 0.
$\alpha = (1, 0)$: $4\lambda^4 x_1^3 - 4\lambda^3 x_2 x_1 \to \ell_1$.
$\alpha = (0, 1)$: $-2\lambda^3 x_1^2 + 2(1 + \varepsilon)\lambda^2 x_2 \to \ell_2$.
$\alpha = (2, 0)$: $12\lambda^4 x_1^2 - 4\lambda^3 x_2 \to \ell_{11}$.

---

[7] Here we get an orthonormal sequence u_n such that $||\Delta_{\Phi_0}^{(0)} u_n||_{L^2(\mathbb{R}^2)} \to 0$,

$\alpha = (0, 2)$: $2(1 + \varepsilon)\lambda^2 \to 0$.

This zero limit is a consequence of the first one which gives $\lambda \to 0$.

$\alpha = (1, 1)$: $-4\lambda^3 x_1 \to \ell_{12}$.

$\alpha = (3, 0)$: $24\lambda^4 x_1 \to \ell_{111}$.

$\alpha = (2, 1)$: $-4\lambda^3 \to 0$.

This zero limit is also a consequence of the first one which gives $\lambda \to 0$.

It is easy to check that the only possible limits must verify

$$\ell_{11} = \ell_{12} = \ell_{111} = 0,$$

for $\varepsilon \neq 0$. Hence $\mathcal{L}_{\Phi_\varepsilon} \cap E_3$ is a set of affine polynomials which have no local minimum and $\Delta_{\Phi_\varepsilon}^{(0)}$ has a compact resolvent.

Spectral Properties of the Witten-Laplacians in Connection with Poincaré Inequalities for Laplace Integrals

11.1 Laplace Integrals and Associated Laplacians

We would like to analyze for a given $\Phi \in C^\infty$ the properties of the measure $\exp -2\Phi \, dx$ where dx is either the canonical measure on a compact riemannian manifold M or the Lebesgue measure on $M = \mathbb{R}^d$. Assuming, at least at the beginning that:

$$Z = \int_M \exp -2\Phi(x) \, dx < +\infty , \tag{11.1}$$

we are interested in Poincaré inequalities, that is in the existence of a constant C_P such that, for all $f \in H^1(M; \exp -2\Phi \, dx)$,

$$\operatorname{var}_\Phi(f) := \langle (f - \langle f \rangle_\Phi)^2 \rangle_\Phi \leq C_P \, \|\nabla f\|^2_{L^2(M, \exp -2\Phi(x) \, dx)} . \tag{11.2}$$

Here,

$$\langle f \rangle_\Phi = Z^{-1} \int_M f(x) \exp -2\Phi(x) \, dx , \tag{11.3}$$

denotes the mean value of f. In the context of the statistical mechanics (see for example [Hel11] for a more detailed presentation), it can be important to control, once the existence of a constant C_P is shown, the dependence of this constant on various parameters (the semi-classical parameter, the dimension, the phase....).

In this context, it has been realized that the Laplacian attached to the Dirichlet form:

$$f \mapsto \|\nabla f\|^2_{L^2(M, \exp -2\Phi \, dx)} \tag{11.4}$$

plays the important role. We call the associated Laplacian $A_\Phi^{(0)}$ the Dirichlet Laplacian.

If, in the case of a compact manifold, the existence of the constant C_P is not a problem.

– The lowest eigenvalue λ_1 of the Dirichlet Laplacian (which has a compact resolvent) is a simple eigenvalue equal to 0, and one can consequently take

$C_P = \frac{1}{\lambda_2}$, where λ_2 is the second one, –

all these questions become more complicate in the case of $M = \mathbb{R}^d$ or more generally when M is not compact.

In this situation, it was recognized by J. Sjöstrand [Sj4], reinterpreting previous results of [HelSj8], that these questions can be more easily understood with the Witten Laplacian approach and this point of view has become popular and has been developed in many contexts by various authors including Bach, Bodineau, Helffer, Jecko, Moeller, Sjöstrand ([BaJeSj, BaMo1, BaMo2, Hel11, HelSj8, Sj4, Sj5]).

The question of the existence of a Poincaré inequality becomes the question of determining if 0 is isolated in the spectrum of the operator $\Delta_\Phi^{(0)}$. We note that the assumption (11.1) is no more needed for stating the problem.

The answer can be a consequence of the property that $\Delta_\Phi^{(0)}$ has a compact resolvent, but this suggests to do more generally the analysis of the essential spectrum of this operator (See for example J. Johnsen [Jo] and the next section).

Moreover, if one forgets the origin of the problem where the assumption (11.1 is naturally made, which has the immediate consequence that "0 is in the spectrum", we can also discuss under which assumption on Φ, this last property is satisfied. This question was analyzed by Helffer-Nier in [HelNi1].

Once these qualitative questions are solved, one can also look for more quantitative control with respect to the various involved parameters and many contributions have been devoted to this problem. We will review some of the semi-classical results in the last chapters.

11.2 Links with the Witten Laplacians

11.2.1 On Poincaré and Brascamp-Lieb Inequalities

Coming from the Laplace integral, one finds more naturally the de Rham complex but with a new adjoint related to the measure $\exp -2\Phi(x)\, dx$. This leads to the operators $A_\Phi^{(j)}$ which are related to the Witten Laplacians by:

$$A_\Phi^{(j)} = \exp \Phi \, \Delta_\Phi^{(j)} \, \exp -\Phi \,. \tag{11.5}$$

Let us show how one can play with the Witten Laplacian in this context. If $A_\Phi^{(0)}$ has 0 as isolated point in the spectrum, then we can find for any $f \in L^2(M, \exp -2\Phi\, dx)$, u such that:

$$f - \langle f \rangle_\Phi = A_\Phi^{(0)} u \,, \tag{11.6}$$

with u orthogonal to the constant functions, that is:

$$\langle u \rangle = 0 \,.$$

Inserting in the formula (11.2) for the variance, we get:

$$\text{var}\,(f) = \langle A_\Phi^{(0)} u \mid f - \langle f \rangle \rangle$$
$$= \langle du \mid df \rangle \, .$$

If we differentiate (11.6), we get:

$$df = dA_\Phi^{(0)} u = A_\Phi^{(1)} du \, .$$

When $A_\Phi^{(1)}$ is invertible (note that under the previous assumption it is at least invertible on the exact forms), we get

$$du = (A_\Phi^{(1)})^{-1} df \, ,$$

and we get the formula:

$$\text{var}\, f = \langle (A_\Phi^{(1)})^{-1} df \mid df \rangle \, . \tag{11.7}$$

This identity has appeared to be very useful in the context of statistical mechanics.
It leads immediately to:

$$|\,\text{var}\,(f)| \le ||(A_\Phi^{(1)})^{-1}||\ ||df||^2 \, , \tag{11.8}$$

where the norms appearing in the right hand side are respectively in $\mathcal{L}(L^2(\exp{-2\Phi}\,dx))$ and $L^2(\exp{-2\Phi}\,dx)$. Although this new inequality is not necessarily optimal, the bottom of the spectrum of $A_\Phi^{(1)}$ can be below the bottom of $\sigma(A_\Phi^{(0)}) \setminus \{0\}$, this inequality can be easier to prove.

The other standard point to mention is that, when Φ is uniformly strictly convex and using the identity:

$$A_\Phi^{(1)} = A_\Phi^{(0)} \otimes I + 2\,\text{Hess}\,\Phi \, , \tag{11.9}$$

we get

$$\text{var}\,(f) \le \frac{1}{2} \langle (\text{Hess}\,\Phi)^{-1} df \mid df \rangle \, , \tag{11.10}$$

which is called the Brascamp-Lieb Inequality.

Proposition 11.1.
If Φ be a C^∞ function such that $2\,\text{Hess}\,\Phi \ge \lambda_1\,\text{Id}$ holds uniformly for some $\lambda_1 > 0$, then the Poincaré inequality (11.2) holds.

11.2.2 Links with Spectra of Higher Order Witten Laplacians

Let us also emphasize, some spectral consequences of this Witten Laplacian approach (which of course was already present in other works by Bakry-Emery) (see [Aetall], [DeuSt], [Jo]). It may be easier to come back to the Witten Laplacians with corresponding Hilbert $L^2(\mathbb{R}^d)$.

Let u be an eigenvector of $\Delta_\Phi^{(0)}$ attached to some eigenvalue $\lambda \neq 0$:

$$\Delta_\Phi^{(0)} u = \lambda u \,,$$

and let us assume for simplicity that $u \in \mathcal{S}(\mathbb{R}^d)$. Then

$$d_\Phi^{(0)} \Delta_\Phi^{(0)} u = \Delta_\Phi^{(1)} d_\Phi^{(0)} u = \lambda d_\Phi^{(0)} u \,. \qquad (11.11)$$

We get then the alternative:

- either the one-form $\omega = d_\Phi^{(0)} u$ is 0 (but this is excluded by $\lambda \neq 0$),
- or it is an eigenvector of $\Delta_\Phi^{(1)}$.

This leads to the inequality:

$$\inf \sigma(\Delta_\Phi^{(1)}) \leq \inf \left(\sigma(\Delta_\Phi^{(0)}) \setminus \{0\} \right) \,. \qquad (11.12)$$

Here we have been rather formal in the description of the argument. J. Johnsen [Jo], using the Weyl Calculus has given rather general criteria under which one can show that $u \in \mathcal{S}(\mathbb{R}^d)$.

It can be shown that we have always equality (11.12) in the one dimensional case, but the inequality may be strict in general.

The next remark is based on discussions with J. Schach Moeller (see [MaMo]) and on the paper of J. Johnsen [Jo]. The idea is to pursue the analysis by comparing the spectra of $\Delta_\Phi^{(1)}$ and $\Delta_\Phi^{(2)}$. If ω is an eigenvector of $\Delta_\Phi^{(1)}$ attached to the eigenvalue λ:

$$\Delta_\Phi^{(1)} \omega = \lambda \omega \,, \qquad (11.13)$$

and let us assume that $\omega \in \mathcal{S}(\mathbb{R}^d; \mathbb{R}^d))$. Then

$$d_\Phi^{(1)} \Delta_\Phi^{(1)} u = \Delta_\Phi^{(2)} d_\Phi^{(1)} \omega = \lambda d_\Phi^{(1)} \omega \,. \qquad (11.14)$$

We have then the alternative that the two-form $\sigma = d_\Phi^{(1)} \omega$ is either 0 or an eigenvector of $\Delta_\Phi^{(2)}$.

The first case in the alternative cannot be treated so easily.

Suppose that we can show that $\omega = d_\Phi^{(0)} u$ with $\langle u \mid e^{-\frac{\Phi}{\hbar}} \rangle = 0$, we get from (11.13):

$$d_\Phi^{(0)} (\Delta_\Phi^{(0)} u - \lambda u) = 0 \,,$$

and consequently, there exists $\gamma \in \mathbb{R}$ such that:

$$\Delta_\Phi^{(0)} u - \lambda u = \gamma \exp - \Phi \,.$$

But taking the scalar product with $\exp - \Phi$, we get: $\gamma = 0$ and u is an eigenvalue of $\Delta_\Phi^{(0)}$.

So if λ is an eigenvalue of $\Delta_\Phi^{(1)}$, then λ is either an eigenvalue of $\Delta_\Phi^{(2)}$ or of $\Delta_\Phi^{(0)}$.

It remains to discuss the existence of u with these properties under the additional assumption that $d_\Phi\omega = 0$. As observed initially by J. Sjöstrand (See [Sj4, Jo, HelNi1]), this can be done if there exists $\delta > 0$ and $C > 0$ such that:

$$x \cdot \nabla\Phi(x) \geq \frac{1}{C}\langle x \rangle^{1+\delta} \text{ for } |x| \geq C . \tag{11.15}$$

This assumption was used for proving the strict positivity of $\Delta_\Phi^{(1)}$. Note that this condition is implied by an assumption of strict convexity at ∞:

$$\liminf_{|x|\to+\infty} \text{Hess}\,\Phi(x) > 0 .$$

We refer to [Jo] and to [HerNi] for further discussions on the subject.

Another question is whether the Poincaré inequality holds when $\sigma(\Delta_\Phi^{(1)}) \subset [\lambda, +\infty)$ for some $\lambda > 0$, without assuming the compactness of the resolvent of $\Delta_\Phi^{(0)}$ nor convexity of Φ.

11.3 Some Necessary and Sufficient Conditions for Polyhomogeneous Potentials

We now summarize some results of [HelNi1], which provide necessary and sufficient conditions for the Poincaré inequality and for the compactness of the resolvent of $\Delta_\Phi^{(0)}$, for a class a C^∞ polyhomogeneous potentials satisfying:

$$\Phi(x) = \sum_{i=1}^N r^{\alpha_i}\varphi_i(\theta) + R(x), \text{ for } x = r\theta , \; r \geq 1 , \; \theta \in \mathbb{S}^{n-1} , \tag{11.16}$$

with $\varphi_i \in C^\infty(\mathbb{S}^{n-1})$, $\alpha_1 > \ldots > \alpha_N > 0$ and $R \in L^\infty(\mathbb{R}^n)$.

11.3.1 Non-negative Polyhomogeneous Potential Near Infinity

Definition 11.2.
With a function Φ like in (11.16), we associate the following sets and functions:

$$\mathcal{Z} := \bigcup_{i=1}^N \varphi_i^{-1}(\{0\}) , \tag{11.17}$$

$$\{\alpha_i > 1\} := \{i \in \{1,\ldots,N\} , \; \alpha_i > 1\}, \tag{11.18}$$

$$\{\alpha_i \geq 1\} := \{i \in \{1,\ldots,N\} , \; \alpha_i \geq 1\} , \tag{11.19}$$

and for all $I \subset \{1,\ldots,N\}$,

$$\mathcal{Z}_I := \bigcap_{i \in I} \varphi_i^{-1}(\{0\}) ,\tag{11.20}$$

$$\Phi_I(x) := \sum_{i \in I} \Phi_i(x) .\tag{11.21}$$

Sufficient conditions for the Poincaré inequality and for the compactness of the resolvent are given by the following result.

Theorem 11.3.
Assume $\Phi \in C^\infty(\mathbb{R}^d)$ of the form (11.16) and let I denote either the set $\{\alpha_i > 1\}$ or the set $\{\alpha_i \geq 1\}$ according to the Definition 11.2. Assume that for $\theta \in \mathbb{S}^{d-1}$, there exists a neighborhood $\mathcal{N}_\theta$ of θ and an index $i_\theta \in I$ such that

$$\varphi_{i_\theta}\Big|_{\mathcal{N}_\theta} > 0 \quad and \quad \forall i \in \{1, \dots, i_\theta - 1\} \ , \ \varphi_i\Big|_{\mathcal{N}_\theta} \geq 0 .\tag{11.22}$$

Case $I = \{\alpha_i > 1\}$:
    The Witten Laplacian $\Delta_\Phi^{(0)}$ has a compact resolvent.
Case $I = \{\alpha_i \geq 1\}$:
    The Poincaré inequality (11.2) holds for some $C_P > 0$.

Corollary 11.4.
Let us assume that all φ_i, $i \in \{1, \dots, N\}$, are non negative.
i) The condition

$$\lim_{|x| \to \infty} \Phi_{\{\alpha_i > 1\}}(x) = +\infty ,\tag{11.23}$$

is sufficient for the compactness of the resolvent of $\Delta_\Phi^{(0)}$.
ii) The condition

$$\liminf_{|x| \to \infty} \Phi_{\{\alpha_i \geq 1\}}(x) > 0 ,\tag{11.24}$$

is sufficient for the validity of the Poincaré inequality (11.2) for some $C_P > 0$.

As stated below the condition $\liminf_{|x| \to \infty} \Phi_{\{\alpha_i \geq 1\}} > 0$ of Corollary 11.4 is necessary.

Theorem 11.5.
Let Φ be given by (11.16) with all φ_i , $i \in \{1, \dots, N\}$, non negative and $\|\partial_r R(r.)\|_{L^\infty(\mathbb{S}^{n-1})} = o(1)$, as $r \to \infty$. If

$$\{\alpha_i \geq 1\} = \emptyset ,$$

or

$$\mathcal{Z}_{\{\alpha_i \geq 1\}} \neq \emptyset ,$$

then $\Delta_\Phi^{(0)}$ has not a compact resolvent and $0 \in \sigma_{ess}(\Delta_\Phi^{(0)})$.

11.3.2 Analysis of the Kernel

By the analysis of the distribution solutions u of $d^{(0)}_{\Phi} u = 0$, the analysis of the kernel amounts to determine if the function $e^{-\Phi}$ belongs to $L^2(\mathbb{R}^n)$. We still consider a potential Φ of the form (11.16). When $N = 1$ and φ_1 is analytic on $\mathbb{S}^{n-1}$, the resolution of singularities technique for Laplace integrals described in [ArGuVar] leads to criteria for $e^{-2\Phi}$ belonging to $L^1(\mathbb{R}^n)$. For $N > 1$ and even with analytic functions φ_i, a general answer does not seem to be known. For the sake of simplicity, we restrict the analysis to the case described below which is generic, if all the φ_i are assumed non negative.

Assumption 11.6.
For $z \in \mathcal{Z}$ and for all $i \in \{1, \ldots, N\}$, there exist $m_i(z) \geq 0$, a neighborhood $\mathcal{N}_i(z) \subset \mathbb{S}^{n-1}$ of z and a constant $c_i > 1$ such that

$$\forall \theta \in \mathcal{N}_i((z)), \; c_i^{-1} |\theta - z|^{m_i(z)} \leq \varphi_i(\theta) \leq c_i |\theta - z|^{m_i(z)} . \tag{11.25}$$

Under this assumption, we associate with Φ the index

$$\mathcal{I}_{\Phi} := \min_{z \in \mathcal{Z}} \max_{i \in \{1, \ldots, N\}} \frac{\alpha_i}{m_i(z)} \in \overline{\mathbb{R}_+} . \tag{11.26}$$

It is not difficult to check

Proposition 11.7.
Let us assume Assumption 11.6. Then

$$e^{-2\Phi} \in L^1(\mathbb{R}^n) , \; \text{if and only if } \mathcal{I}_{\Phi} > \frac{n}{n-1} .$$

11.3.3 Non-positive Polyhomogeneous Potential Near Infinity

Here the situation is even simpler because for the $\varphi_i \leq 0$, the term $-\Delta\Phi$ is nonnegative. It suffices to get a lower bound of $-\Delta + |\nabla\Phi|^2$ by following the Kohn method adapted by Helffer-Morame [HelMo] and presented in Section 3.3. Note that in this case 0 cannot be an eigenvalue of $\Delta^{(0)}_{\Phi}$ and the compactness of the resolvent implies $\Delta^{(0)}_{\Phi} > 0$.

Let A be the integer part of the maximum degree α_1 in (11.16). We associate with Φ the function

$$q(x) := \sum_{1 \leq |\beta| \leq A} \left| \partial_x^{\beta} \Phi(x) \right| . \tag{11.27}$$

The exact assumptions are

Assumption 11.8.
The function $\Phi \in C^\infty(\mathbb{R}^n)$ has the form (11.16) with $\lim_{|x|\to\infty} q(x) = +\infty$. Let us assume also that, for all $\theta \in \mathbb{S}^{d-1}$,

either:

   there exist i_θ such that $\alpha_{i_\theta} > 1$ and a neighborhood $\mathcal{N}_\theta$ of θ such that

$$\varphi_{i_\theta}\big|_{\mathcal{N}_\theta} < 0\,, \quad \forall i \in \{1,\ldots,i_\theta\}\,,\ \varphi_i\big|_{\mathcal{N}_\theta} \le 0\,;$$

or:

   there exists a neighborhood $\mathcal{N}_\theta$ of θ such that

$$\forall i \text{ with } \alpha_i > 2,\ \varphi_i\big|_{\mathcal{N}_\theta} \le 0 \text{ and } \Delta_\theta \varphi_i\big|_{\mathcal{N}_\theta} \le 0\,.$$

The result is:

Theorem 11.9.
Under Assumption 11.8, the operator $\Delta_\Phi^{(0)}$ has a compact resolvent.

11.4 Applications in the Polynomial Case

11.4.1 Main Result

We now show how the results of Sections 11.3 and 10.4 give a complete answer to the questions raised in Section 11.1 at least for polynomial potentials (with the above sign conditions). First of all, the results of the previous section give an alternative and more direct characterization (in comparison with Theorem 10.16) for some specific class of polynomials.

Theorem 11.10.
Let $\Phi \in \mathbb{R}[X_1, \ldots, X_d]$ be a polynomial potential.

i) If Φ is a sum of non negative monomials, then we have:

$$(11.2) \Leftrightarrow \left(\lim_{|x|\to\infty} \Phi(x) = +\infty\right) \Leftrightarrow \left((1 + \Delta_\Phi^{(0)})^{-1} \text{ compact}\right).$$

ii) If Φ is a sum of non positive monomials, then $\Delta_\Phi^{(0)}$ has a compact resolvent if and only if $q(x) := \sum_{|\alpha|>0} |D_x^\alpha \Phi(x)| \to +\infty$.

We refer to [HelNi1] for a detailed proof but let us just mention the three simple properties which are involved in the proof.

   1. The first point is that one can localize using a partition of unity (see (12.11)) the proof in suitable cones.

2. The second point is that the existence of a Poincaré inequality or the compactness of the resolvent does not depend on the addition of a bounded function (see [DeuSt]).

3. The last point is that, for any decomposition of $\Phi = \Phi_1 + \Phi_2$, one has the identity:

$$\Delta_\Phi^{(0)} = \Delta_{\Phi_1}^{(0)} + 2\nabla\Phi_1 \cdot \nabla\Phi_2 + |\nabla\Phi_2|^2 - \Delta\Phi_2 . \tag{11.28}$$

This is particularly useful when we can find a decomposition such that:
- $\nabla\Phi_1 \cdot \nabla\Phi_2 \geq 0$,
- $|\nabla\Phi_2(x)|^2 - \Delta\Phi_2(x)$ tends to $+\infty$ as $|x| \to +\infty$.

The most typical example is $\Phi = x_1^2 x_2^2 + \epsilon(x_1^2 + x_2^2)$ with $\epsilon > 0$, which will be discussed below.

11.4.2 Examples

Here is a list of examples which show various possibilities.

11.4.2.a: $\Phi = x_1^2 x_2^2$ in $\mathbb{R}^2$.

We have $\mathcal{Z}_{\{\alpha_i \geq 1\}} \neq \emptyset$ and 0 belongs to the essential spectrum of $\Delta_\Phi^{(0)}$. The Poincaré inequality (11.2) does not hold. Assumption 11.6 is satisfied and the index $\mathcal{I}_\varphi$ equals $2 = \frac{2}{2-1}$. Thus $e^{-2\Phi}$ is not in $L^1(\mathbb{R}^d)$ and 0 is not an eigenvalue of $\Delta_\Phi^{(0)}$.

11.4.2.b: $\Phi = x_1^2 x_2^2 (x_1^2 + x_2^2)$ in $\mathbb{R}^2$.

We have $\mathcal{Z}_{\{\alpha_i \geq 1\}} \neq \emptyset$ and $\mathcal{I}_\Phi$ equals $3 > 2$. Thus, 0 is an eigenvalue contained in the essential spectrum.

11.4.2.c: $\Phi = x_1^2 x_2^2 x_3^2$ in $\mathbb{R}^3$.

We have $\mathcal{Z}_{\{\alpha_i \geq 1\}} \neq \emptyset$ and the Poincaré inequality is not satisfied. This case does not satisfy Assumption 11.6 but integration with respect to $x_3 \in \mathbb{R}$ shows that the function $e^{-\Phi}$ is not in $L^2(\mathbb{R}^3)$.

11.4.2.d: $\Phi = (x_1^2 + x_2^2)(x_2^2 + x_3^2) + (x_1^2 + x_3^2)$ in $\mathbb{R}^3$.

Then the set $\mathcal{Z}_{\{\alpha_i > 1\}} = \emptyset$ and $\Delta_\Phi^{(0)}$ has a compact resolvent.

11.4.2.e: $\Phi = (1 + |x^2|)^{1/2}$ in $\mathbb{R}^d$.

The function $e^{-2\Phi}$ belongs to $L^1(\mathbb{R}^d)$. The Poincaré inequality is satisfied because $\mathcal{Z}_{\{\alpha_i \geq 1\}} = \emptyset$. But since $V = |\nabla\Phi|^2 - \Delta\Phi$ belongs to $L^\infty(\mathbb{R}^d)$, the resolvent of $\Delta_\Phi^{(0)}$ is not compact.

11.4.2.f: $\Phi_\varepsilon = x_1^2 x_2^2 + \varepsilon(x_1^2 + x_2^2)$ in $\mathbb{R}^2$.

The Witten Laplacian $\Delta_{\Phi_\varepsilon}^{(0)}$ has a compact resolvent for any $\varepsilon > 0$. This was also observed by a different approach in Proposition 10.21. Hence the $\Delta_{\Phi_\varepsilon}^{(0)}$ does not converge to $\Delta_{\Phi_0}^{(0)}$ in the norm resolvent sense as $\varepsilon \to 0$. Note also that, for any $\varepsilon > 0$, this potential is far from being convex since the determinant of its Hessian

$$\begin{pmatrix} 2x_2^2 + \varepsilon & 4x_1x_2 \\ 4x_1x_2 & 2x_1^2 + \varepsilon \end{pmatrix}$$

is equal to $(2t^2 + \varepsilon)^2 - 16t^4$ on the lines $x_1 = \pm x_2 = t$.

11.4.2.**g**: $\Phi_\varepsilon = (x_1^2 - x_2)^2 + \varepsilon x_2^2$ in $\mathbb{R}^2$ with $\varepsilon \in \mathbb{R}$.
This potential is equal to

$$\Phi_\varepsilon = x_1^4 - 2x_1^2x_2 + (1+\varepsilon)x_2^2$$

and it does not satisfy the assumptions of Theorem 11.10, Theorem 11.5, Theorem 11.3 and Theorem 11.7 since the term with total degree 3 is negative for $x_1 \neq 0$, $x_2 > 0$. As mentioned in [HelNi1], it is possible to get the compactness of the resolvent via a simple comparison argument (see 3. after Theorem 11.10) for $\varepsilon > 1/8$. The general case was analyzed in Section 10.4.

11.5 About the Poincaré Inequality for an Homogeneous Potential

Here we make the connection between the results of Sections 10.4 and 11.3. We consider the simple case of an homogeneous potential near ∞ without sign condition:

$$\forall x \in \mathbb{R}^n, |x| \geq 1, \quad \Phi(x) = |x|^m \, \varphi(\frac{x}{|x|}) . \tag{11.29}$$

With the homogeneity degree m, we associate the integer

$$\widehat{m} = \max\{\mu \in \mathbb{N}, \mu < m\} . \tag{11.30}$$

We shall provide here various necessary and sufficient conditions for the compactness of the resolvent of $\Delta_\Phi^{(0)}$. The sufficient conditions will rely on maximal or non-maximal microhypoellipticity of associated complex differential systems and the comparison of the two cases will be done. When φ is a Morse function, we have necessary and sufficient conditions which depend on m. We end this analysis with a remark on the links between the topology of Φ at infinity (that is of φ) and the compactness of the resolvent.

11.5.1 Necessary Conditions

A necessary condition is given in the next proposition.

Proposition 11.11.
Assume that the $\Phi \in C^\infty(\mathbb{R}^n)$ satisfies (11.29). If the Witten Laplacian $\Delta_\Phi^{(0)}$ has a compact resolvent, then

i) $m > 1$;

ii) φ *does not vanish at order* $\widehat{m} + 1$, $\sum_{|\alpha| \leq \widehat{m}} |\partial_\theta^\alpha \varphi(\theta)| > 0$;

iii) *There is no pair* $K \subset U$ *with* $K \subset \varphi^{-1}(\{0\})$ *compact,* $K \neq \emptyset$ *and* $U \subset \mathbb{S}^{n-1}$ *open, such that*

$$\forall \theta \in U \setminus K, \quad \varphi(\theta) > 0 .$$

Proof of i)

The condition $m > 1$ is obviously[1] necessary and we will restrict our attention to this case.

Proof of ii)

Let us assume that φ vanishes at θ_0 at order $\widehat{m} + 1$. Since φ is a C^∞ function there exist a neighborhood V_{θ_0} and a constant C_{θ_0} such that

$$\forall \theta \in V_{\theta_0}, \quad -C_{\theta_0} |\theta - \theta_0|^{\widehat{m}+1} \leq \varphi(\theta) \leq C_{\theta_0} |\theta - \theta_0|^{\widehat{m}+1} .$$

For a function $\chi_1 \in C_0^\infty(\mathbb{R})$, $\chi_1 = 1$ in a neighborhood of 0, $\chi_1 \geq 0$, there exist two constants $R_1 > 0$ and $C_1 > 0$, so that the integral

$$I_{\chi_1}(r) = \int_{\mathbb{S}^{n-1}} \left(\chi_1(r^{\frac{m}{\widehat{m}+1}} |\theta - \theta_0|) \right)^2 e^{-2r^m \varphi(\theta)} \, d\theta$$

satisfies

$$\forall r \geq R_1, \quad C_1^{-1} r^{-\frac{(n-1)m}{\widehat{m}+1}} \leq I_{\chi_1}(r) \leq C_1 r^{-\frac{(n-1)m}{\widehat{m}+1}} . \tag{11.31}$$

Let χ_2 be a $C^\infty(]1/2, 3[)$ function, such that $\chi_2 = 1$ on $[1, 2]$, $\chi_2 \geq 0$. We introduce the family of functions

$$u_\varepsilon(x) = u_\varepsilon(r\theta) = \chi_2(\varepsilon r) \chi_1(r^{\frac{m}{\widehat{m}+1}} |\theta - \theta_0|) e^{-r^m \varphi(\theta)} , \text{ for } \varepsilon > 0 ,$$

written with $r = |x|$ and $\theta = \frac{x}{|x|}$. For $u(x) = a(x) e^{-\Phi(x)}$, the Φ-differential $d_\Phi u = e^{-\Phi} da$ equals $\begin{pmatrix} \partial_r a \\ \frac{1}{r} \partial_\theta a \end{pmatrix} e^{-\Phi}$ in a polar basis. Hence the radial component of $d_\Phi u_\varepsilon$ is estimated by

$$\left| e^{-\Phi} \partial_r (e^\Phi u_\varepsilon) \right| \leq C(r^{-1} + \varepsilon) \gamma_\varepsilon(r, \theta) \leq C \, \varepsilon \gamma_\varepsilon(r, \theta) ,$$

and the angular component by

$$\left| e^{-\Phi} \frac{1}{r} \partial_\theta (e^\Phi u_\varepsilon) \right| \leq C \, r^{\frac{m}{\widehat{m}+1} - 1} \gamma_\varepsilon(r, \theta) \leq C \varepsilon^{-\frac{m - \widehat{m} - 1}{m}} \gamma_\varepsilon(r, \theta) ,$$

where γ_ε is given by $\gamma_\varepsilon(r, \theta) = 1_{[\frac{1}{2\varepsilon}, \frac{3}{\varepsilon}]}(r) \tilde{\chi}_1(r^{\frac{m}{\widehat{m}+1}} |\theta - \theta_0|)$, for some cut-off function $\tilde{\chi}_1$ which equals 1 on $\text{supp} \, \chi_1$.

Thus we have

---

[1] If not, $\Delta_\Phi^{(0)}$ is a bounded perturbation of $-\Delta$.

$$\left|e^{-\Phi}\partial_r(e^{\Phi}u_\varepsilon)\right|^2 + \left|e^{-\Phi}\frac{1}{r}\partial_\theta(e^{\Phi}u_\varepsilon)\right|^2 \leq C\,\varepsilon^{-2\frac{m-\widehat{m}-1}{m}}\,\gamma_\varepsilon(r,\theta)\ .$$

After integrating with respect to θ with the upper bound (11.31) applied to $I_{\tilde{\chi}_1}(r)$, we obtain

$$\int_{\mathbb{S}^{n-1}} \left|e^{-\Phi}\partial_r(e^{\Phi}u_\varepsilon)\right|^2 + \left|e^{-\Phi}\frac{1}{r}\partial_\theta(e^{\Phi}u_\varepsilon)\right|^2 \leq C\varepsilon^{-2\frac{m-\widehat{m}-1}{m}}1_{[\frac{1}{2\varepsilon},\frac{3}{\varepsilon}]}(r)\varepsilon^{\frac{m(n-1)}{\widehat{m}+1}}\ .$$

Meanwhile the lower bound (11.31) applied to $I_{\tilde{\chi}_1}$ gives

$$\int_{\mathbb{S}^{n-1}} |u_\varepsilon|^2\ d\theta \geq C^{-1}1_{[\frac{1}{\varepsilon},\frac{2}{\varepsilon}]}(r)\varepsilon^{\frac{m(n-1)}{\widehat{m}+1}}\ .$$

After integrating with $r^{n-1}dr$, there exist ε_0 and $C > 0$ such that

$$\forall\varepsilon\in]0,\varepsilon_0],\ \frac{\langle u_\varepsilon\mid\Delta_\Phi^{(0)}u_\varepsilon\rangle}{\|u_\varepsilon\|^2} = \frac{\|d_\Phi u_\varepsilon\|^2}{\|u_\varepsilon\|^2} \leq C\varepsilon^{2\frac{\widehat{m}+1-m}{m}}\ .$$

By taking $\varepsilon_n = 8^{-n}$, we obtain a sequence of functions u_{ε_n} with disjoint supports so that the Rayleigh quotient $\frac{\langle u_{\varepsilon_n}\mid\Delta_\Phi^{(0)}u_{\varepsilon_n}\rangle}{\|u_{\varepsilon_n}\|^2}$ is bounded as $n\to\infty$. It is a Weyl sequence which gives the boundedness of $\min\sigma_{ess}(\Delta_\Phi^{(0)})$ (this minimum is 0 if $m\notin\mathbb{N}$).

Proof of iii)

After a localization in θ independent of r, the construction of a Weyl sequence reduces to the case where $\varphi\geq 0$ on $\mathbb{S}^{n-1}$ and vanishes on K. This case was considered in [HelNi1] and we refer to it for the proof.   ∎

11.5.2 Sufficient Conditions

We now consider sufficient conditions. We will see that the necessary condition iii) of Proposition 11.11, which says that φ admits no 0-valued minimum has to be strengthened. Indeed the compactness of the resolvent of $\Delta_\Phi^{(0)}$, with $\Phi(r\theta) = |x|^m\varphi(\frac{x}{|x|})$ is related to the microhypoellipticity of the system $\partial_{\theta_j} + (\partial_{\theta_j}\varphi(\theta))D_t$ near the point $(\theta_0,t_0,\hat{\theta},\tau = +1)$ in $(\mathbb{S}^{n-1}\times\mathbb{R}_t)\times(\mathbb{R}^n\setminus\{0\})$, where θ_0 is a zero of φ. A condition related to Remark 10.6 permits maximal hypoellipticity arguments and leads to the compactness of the resolvent for $\Delta_\Phi^{(0)}$. Meanwhile the condition that φ is analytic with no 0-valued minimum, which leads to microhypoellipticity properties, will not be sufficient in general.

We start with a lemma which will be used in the different cases. First we introduce for $c > 1$, the shell:

$$\mathcal{S}_c = \left\{x\in\mathbb{R}^d,\ c^{-1} < |x| < c\right\}\ ,$$

with closure $\overline{\mathcal{S}}_c$. We will use the notation $L^{2,s}$ for the weighted L^2 space $L^2(\mathbb{R}^n,\langle x\rangle^{2s}\ dx)$ with the corresponding norm $\|u\|_{L^{2,s}} = \|\langle x\rangle^s u\|$.

Lemma 11.12.
Let $\Phi \in C^\infty(\mathbb{R}^n)$ be of the form (11.29), $m > 1$, and such that φ does not vanish at order $\hat{m} + 1$. If for some $c > 1$ and $\mu \geq 1$, the function Φ is homogeneous in $\{|x| \geq c^{-1}\}$ and the semiclassical Witten Laplacian $\Delta^{(0)}_{\Phi,h}$ is $(1 - \frac{1}{\mu})$-subelliptic in S_c, according to Definition 10.2, then the Witten Laplacian $\Delta^{(0)}_{\Phi}$ ($h = 1$) is bounded from below by $C^{-1}\langle x \rangle^{2(\frac{m}{\mu}-1)} - C$ and has a compact resolvent if $\frac{m}{\mu} > 1$.

Let $x \mapsto \chi_0(x)^2 + \sum_{j=1}^\infty \chi^2(c^{-j}x)$ be a c-adic partition of unity on $[0, \infty)$, with $\chi_0 \in C_0^\infty([0, \infty)$ equal to 1 in a neighborhood of 0 and $\chi \in C_0^\infty(]c^{-1}, c[)$. For $u \in C_0^\infty(\mathbb{R}^d)$, we set $u_0 = \chi_0(|x|)u$ and for $j \geq 1$, $u_j = \chi(c^{-j}|x|)u$. Then the norm of $\| \; \|_{L^{2,s}}$ is equivalent to

$$\|u\|_{L^{2,s}}^2 = \|\langle x \rangle^s u\|^2 \sim \sum_{j \in \mathbb{N}} c^{2js} \|u_j\|^2 \; ,$$

with constants determined by $s \in \mathbb{R}$ and the partition of unity. We write again

$$\Delta^{(0)}_{\Phi} = \chi_0 \Delta^{(0)}_{\Phi} \chi_0 + \sum_{j=1}^\infty \chi(c^{-j}|x|)\Delta^{(0)}_{\Phi}\chi(c^{-j}|x|)$$

$$- |\chi_0'(|x|)|^2 - \sum_{j=1}^\infty c^{-2j} \left| \chi'(c^{-j}|x|) \right|^2 .$$

From this we get

$$\langle u \mid \Delta^{(0)}_{\Phi} u \rangle \geq \sum_{j=1}^\infty \langle u_j \mid \Delta^{(0)}_{\Phi} u_j \rangle - C \sum_{j=0}^\infty c^{-2j} \|u_j\|^2 .$$

For the first sum, we set $v_j(x) = c^{nj/2}u_j(c^j x)$, so that all the v_j's belong to $C_0^\infty(S_c)$ with $\|v_j\| = \|u_j\|$. Due to the homogeneity of Φ, we have

$$\langle u_j \mid \Delta^{(0)}_{\Phi} u_j \rangle = c^{j(2m-2)} \langle v_j \mid \Delta^{(0)}_{\Phi,c^{-mj}} v_j \rangle .$$

By referring to Definition 10.2, we assumed that the semiclassical Witten Laplacian $\Delta^{(0)}_{\Phi,h}$ satisfies

$$\forall v \in C_0^\infty(S_c), \quad \langle \Delta^{(0)}_{\Phi,h} v \mid v \rangle \geq C^{-1} h^{2(1-\frac{1}{\mu})} \|v\|^2 \; ,$$

uniformly for $h \in (0, h_0]$, $h_0 > 0$. Hence we, we get for $j_0 > 0$ large enough:

$$\forall j \geq j_0, \quad \langle v_j \mid \Delta^{(0)}_{\Phi,c^{-mj}} v_j \rangle \geq C^{-1} c^{-mj(2-\frac{2}{\mu})} \|v_j\|^2 ,$$

which means

$$\forall j \geq j_0, \quad \langle u_j \mid \Delta^{(0)}_{\Phi} u_j \rangle \geq C^{-1} c^{(\frac{2m}{\mu}-2)j} \|u_j\|^2 .$$

Summing over j the previous estimates, we get

$$\forall u \in C_0^\infty(\mathbb{R}^n), \quad \langle u \mid \Delta_\Phi^{(0)} u \rangle \geq C^{-1} \|u\|_{L^{2,\frac{m}{\mu}-1}}^2 - C \|u\|^2 .$$

This yields $\Delta_\Phi^{(0)} \geq C^{-1} \langle x \rangle^{2(\frac{m}{\mu}-1)} - C$ and $\Delta_\Phi^{(0)}$ has a compact resolvent as soon as $\frac{m}{\mu} > 1$. ∎

Remark 11.13.
For $\frac{m}{\mu} > 1$ we proved that the domain $D(\Delta_\Phi^{(0)})$ is contained in $L^{2,\frac{m}{\mu}-1}$. One can show similarly, for any $s \in \mathbb{R}$, the estimate

$$\forall u \in C_0^\infty(\mathbb{R}^n), \quad \|u\|_{L^{2,s+\frac{m}{\mu}-1}}^2 \leq C_s \left[\left\| \Delta_\Phi^{(0)} u \right\|_{L^{2,s}}^2 + \|u\|_{L^{2,s}}^2 \right],$$

which says that the resolvent sends $L^{2,s}$ into $L^{2,s+\frac{m}{\mu}-1}$.

Here is a sufficient condition for the compactness of the resolvent of $\Delta_\Phi^{(0)}$ which relies on maximal hypoellipticity.

Proposition 11.14.
Assume that $\Phi \in C^\infty(\mathbb{R}^n)$ has the form (11.29), $m > 1$, and satisfies
(1) φ does not vanish at order $\widehat{m} + 1$
and
(2) For all $\theta_0 \in \varphi^{-1}(\{0\})$, there exist a neighborhood V_{θ_0} of θ_0 and two constants $d_{\theta_0} > 0$ and $c_{\theta_0} > 0$, such that, for all $d \in]0, d_{\theta_0}]$, for all $\theta_1 \in V_{\theta_0}$,

$$\inf_{|\theta - \theta_1| \leq d} (\varphi(\theta) - \varphi(\theta_1)) \leq -c_{\theta_0} \sup_{|\theta - \theta_1| \leq d} |\varphi(\theta) - \varphi(\theta_1)| . \tag{11.32}$$

Then the Witten Laplacian $\Delta_\Phi^{(0)}$ has a compact resolvent.

Remark 11.15.
In dimension $n = 2$, the sphere $\mathbb{S}^{n-1}$ is one dimensional and the condition (11.32) is equivalent to the absence of 0-valued local minimum.

Indeed it will be a straightforward application of Lemma 11.12 once we have checked that the semiclassical Witten Laplacian $\Delta_{\Phi,h}^{(0)}$ is $(1 - \frac{1}{\widehat{m}})$-subelliptic in any shell S_c, $c > 1$. We fix $c > 1$ and first work in the shell S_{2c} (by assuming that Φ is homogeneous in $\{|x| \geq (2c)^{-1}\}$) For x_0 in S_{2c} one can find $d_{x_0} > 0$, $c_{x_0} > 0$ and a neighborhood V_{x_0} such that:

$$\forall d \in]0, d_{x_0}], \forall x_1 \in V_{x_0}, \quad \inf_{|x-x_1| \leq d} (\Phi(x) - \Phi(x_1)) \leq -c_{x_0} \sup_{|x-x_1| \leq d} |\Phi(x) - \Phi(x_1)| .$$

Indeed there are two cases:

1. Either $\frac{x_0}{|x_0|}$ does not belong to $\varphi^{-1}(\{0\})$ and $\partial_r \Phi(x_0) \neq 0$ so that $|\nabla \Phi|$ is uniformly bounded from below in a small enough neighborhood of x_0.
2. Or $\frac{x_0}{|x_0|}$ belongs to $\varphi^{-1}(\{0\})$ and it is a consequence of the condition (11.32) combined with the homogeneity of Φ in $(2c)^{-1} < r < 2c$.

According to Theorem 10.5, this leads to the semiclassical estimate

$$\forall u \in C_0^\infty(V_{x_0}), \; h^{2-\frac{2}{\mu_{x_0}}} \|u\|^2 \leq C_{x_0} \langle u \mid \Delta_{\Phi,h}^{(0)} u \rangle, \tag{11.33}$$

for some constant $C_{x_0} > 0$ and with $\mu_{x_0} \leq \widehat{m}$ due to the non vanishing of φ at order $\widehat{m} + 1$. Here μ_{x_0} which replaces the r of Theorem 10.5 is the vanishing order of φ at $\frac{x_0}{|x_0|}$. This result holds for any $x_0 \in S_{2c}$ and in particular for any $x_0 \in \overline{S_c}$. For any partition of unity $\sum_{i=1}^N \chi_i^2 = 1$ on $\overline{S_c}$, with $\chi_i \in C_0^\infty(\mathbb{R}^n)$, we have

$$\forall u \in C_0^\infty(S_c), \quad \langle u \mid \Delta_{\Phi,h}^{(0)} u \rangle - \sum_{i=1}^N \langle \chi_i u \mid \Delta_{\Phi,h} \chi_i u \rangle = -h^2 \sum_{i=1}^N \langle u \mid |\nabla \chi_i|^2 u \rangle.$$

The compactness of $\overline{S_c}$ then leads to

$$\forall u \in C_0^\infty, \; h^{2-\frac{2}{\mu}} \|u\|^2 \leq C_c \langle u \mid \Delta_{\Phi,h}^{(0)} u \rangle, \tag{11.34}$$

for $h \in]0, h_c]$, with the constants $C_c > 0$ and $h_c > 0$ depending on c and where $\mu \in \mathbb{N}$, $\mu \leq \widehat{m} < m$, is the maximal vanishing order of φ. ∎

11.5.3 The Analytic Case

If the condition iii) (with i) and ii)) of Proposition 11.11 is satisfied and if the function φ is real analytic, then Φ is a real analytic function in the shell S_c, $c > 1$, without any minimum. Compared to Proposition 11.14, the analyticity property and the absence of 0-valued local minimum for φ are weaker assumptions. In Proposition 11.14, the condition (11.32) on φ, which gives the maximal microhypoellipticity, is semilocal in the sense that a uniform estimate has to be satisfied in the neighborhood of any point where φ vanishes. This condition adds to the absence of 0-valued minimum a control of higher order derivatives in the neighborhood of any point of $\varphi^{-1}(\{0\})$. Meanwhile, the condition that a real analytic φ has no 0-valued minimum does not contain enough information on the behaviour of φ around $\varphi^{-1}(\{0\})$.

Although the condition iii) of Proposition 11.11 leads in the analytic framework according to Maire in [Mai1, Mai4] to microhypoellipticity properties of the system $\partial_{\theta_j} + (\partial_{\theta_j}\varphi(\theta))D_t$, it is not sufficient to ensure, after adding the conditions i) and ii) of Proposition 11.11, the compactness of the resolvent of $\Delta_\Phi^{(0)}$, contrary to the maximal microhypoelliptic framework (Proposition 11.14). Actually the maximal microhypoellipticity which is of algebraic nature is more robust than the microhypoellipticity studied by Maire.

We will even give a more accurate version of these remarks by considering an example similar to (10.22).

An interesting example.

Take in $\mathbb{R}^3$ a function Φ of the form (11.29), with $m > 2$ such that $\sum_{|\alpha| \leq 1} |\partial_\theta^\alpha \varphi(\theta)| \neq 0$ except at the point θ_0 around which φ is analytic. Assume moreover that, in (normal) local coordinates $\theta = (\theta_1, \theta_2)$ with $\theta_0 = (0, 0)$ the function φ is given by

$$\varphi(\theta_1, \theta_2) = \theta_1^{2\ell+1} - \theta_1 \theta_2^2, \quad \ell > 1.$$

Let $c > 1$. Around any $x \in \mathcal{S}_c$ such that $\frac{x}{|x|} \neq \theta_0$, the condition (11.32) is satisfied for $\nabla \Phi(x) \neq 0$ and the semiclassical Witten Laplacian $\Delta_{\Phi,h}^{(0)}$ is $1/2$-subelliptic.

If the result stated in [Mai4] holds for the L^2-subellipticity (See the discussion of Section 10.2 around Conjecture 10.8 and the footnote 3) this implies that $\Delta_{\Phi,h}^{(0)}$ is $\frac{2\ell}{2\ell+1}$-subelliptic in a neighborhood of any point $x \in \mathcal{S}_c$ such that $\frac{x}{|x|} = \theta_0$. Hence we have $\mu \geq 2\ell + 1$ and $\frac{m}{\mu}$ is larger than 1 if $m > 2\ell + 1$. According to Lemma 11.12 a positive answer to Conjecture 10.8 in this case implies that $\Delta_\Phi^{(0)}$ has a compact resolvent if $m > 2\ell + 1$.

Note that the function φ satisfies $\sum_{|\alpha| \leq 3} |\partial_\theta^\alpha \varphi(\theta)| \neq 0$ for all $\theta \in S^{d-1}$. Under the condition (11.32) (which is not true here at $\theta = \theta_0$) the compactness of the resolvent would hold as soon as $m > 3$.

We now check that the condition $m > 2\ell + 1$ is optimal.

Proposition 11.16. *With the previous choice of the function φ, the Witten Laplacian $\Delta_\Phi^{(0)}$, with $\Phi = |x|^m \varphi(\frac{x}{|x|})$ for $|x| \geq 1$, does not have a compact resolvent without the additional condition $m > 2\ell + 1$. Moreover $0 \in \sigma_{ess}\left(\Delta_\Phi^{(0)}\right)$ if $m < 2\ell + 1$.*

In particular if $\ell > 1$, this says that $\Delta_\Phi^{(0)}$ does not have a compact resolvent when $m \in (3, 2\ell + 1]$ although the conditions i), ii) and iii) of Proposition 11.11 are satisfied.

Proof: Let χ_1 be a $\mathcal{C}^\infty(\mathbb{R}^2)$ function such that $0 \leq \chi_1 \leq 1$ with $\operatorname{supp} \chi_1 \subset \{|\theta| < 1\}$ and $\chi_1(\theta) \equiv 1$ for $|\theta| \leq 1/2$. Let χ_2 be in $\mathcal{C}^\infty(]c^{-1}, c[)$. We set

$$\chi_1^h(\theta) = \chi_1 \left(\frac{\theta_1 + 2h^{\frac{1}{2\ell+1}}}{h^{\frac{1}{2\ell+1}}}, \frac{\theta_2}{h^{\frac{1}{2\ell+1}}} \right), \quad \chi^h(x) = \chi_1^h(\frac{x}{|x|}) \chi_2(|x|),$$

and $v_h = \chi^h(x) e^{-\Phi(x)/h}$.

We have

$$\frac{\left\langle \Delta_{\Phi,h}^{(0)} v_h \mid v_h \right\rangle}{\|v_h\|^2} = \frac{\|(h\nabla \chi^h) e^{-\Phi(x)/h}\|^2}{\|\chi^h e^{-\Phi(x)/h}\|} .$$

We are led to estimate the quotient

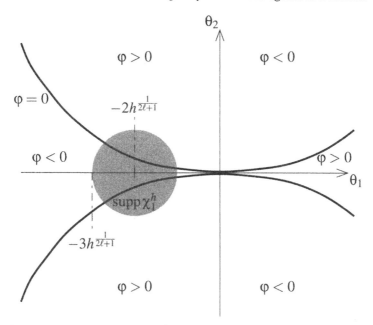

Fig. 11.1. Position of $\operatorname{supp}\chi_1^h$ with respect to the level curves of φ.

$$\frac{\int \left|\nabla_\theta \chi_1^h(\theta)\right|^2 e^{-2\lambda\varphi(\theta)/h}\,\widetilde{d\theta}}{\int \left|\chi_1^h(\theta)\right|^2 e^{-2\lambda\varphi(\theta)/h}\,\widetilde{d\theta}}\ ,\tag{11.35}$$

with $\lambda \in [c^{-m}, c^m]$, $\widetilde{d\theta} = \alpha(\theta)\,d\theta$, $\alpha(0) = 1$.

The support of χ_1^h is contained in $\theta_1 < 0$ (indeed in the ball centered at $(-2h^{1/(2\ell+1)}, 0)$ with radius $h^{1/(2\ell+1)}$) so that

$$\forall \theta \in \operatorname{supp}\chi_1^h,\ (\varphi(\theta) < 0) \Leftrightarrow \left\{|\theta_2| \leq |\theta_1|^\ell\right\}\ .$$

By taking the constant $C > 0$ large enough, the quotient behaves like

$$\frac{\int_{|\theta_2|\leq C|\theta_1|^\ell} \left|\nabla_\theta \chi_1^h(\theta)\right|^2 e^{-2\lambda\varphi(\theta)/h}\,\widetilde{d\theta}}{\int_{|\theta_2|\leq C|\theta_1|^\ell} \left|\chi_1^h(\theta)\right|^2 e^{-2\lambda\varphi(\theta)/h}\,\widetilde{d\theta}}\ .$$

In $\operatorname{supp}\chi_1^h \cap \left\{|\theta_2| \leq C\,|\theta_1|^\ell\right\}$ the estimate $\varphi(\theta) \geq -C'h$ yields

$$\frac{\int \left|\nabla_\theta \chi_1^h(\theta)\right|^2 e^{-2\lambda\varphi(\theta)/h}\,\widetilde{d\theta}}{\int \left|\chi_1^h(\theta)\right|^2 e^{-2\lambda\varphi(\theta)/h}\,\widetilde{d\theta}} = \mathcal{O}(h^{-2/(2\ell+1)})\ .$$

We obtain

$$\left\langle \Delta_{\Phi,h}^{(0)} v_h \mid v_h \right\rangle \leq C'' h^{2-2/(2\ell+1)} \|v_h\|^2\ ,$$

and $\Delta_{\Phi,h}^{(0)}$ cannot be $(1 - 1/\mu)$-subelliptic with $\mu < 2\ell + 1$.

Finally a sequence $u_k = h_k^{n/2} v_{h_k}(h_k x)$, with $(h_k)_{k\in\mathbb{N}}$ will be a Weyl sequence for $\Delta_\Phi^{(0)}$ so that the resolvent of $\Delta_\Phi^{(0)}$ is not compact for $m \leq 2\ell + 1$ and $0 \in \sigma_{ess}\left(\Delta_\Phi^{(0)}\right)$ if $m < 2\ell + 1$. ∎

11.5.4 Homotopy Properties

We end this chapter by describing some homotopy properties of Witten Laplacians on zero forms associated with homogeneous functions of degree $m > 1$. If one studies the full Witten complex given by $d_\Phi = e^{-\Phi} d e^\Phi$, the kernel $\text{Ker}\,\Delta_\Phi^{(0)}$ is the homology group of order 0. It is equal to $\mathbb{C}\,e^{-\Phi}$ when $\varphi > 0$ and to $\{0\}$ when $\varphi(\theta_0) < 0$ for some θ_0. This suggests that one cannot go from the case $\varphi > 0$ on $\mathbb{S}^{n-1}$ to the case $\varphi(\theta_0) < 0$ for some θ_0 while keeping the compactness of the resolvent. We start with a definition.

Definition 11.17.
Two Witten Laplacians $\Delta_{\Phi_0}^{(0)}$, $\Delta_{\Phi_1}^{(0)}$, with $\Phi_i = r^m \varphi_i(\theta)$ for $r \geq 1$ and $\varphi_i \in C^\infty(\mathbb{S}^{n-1})$ are said homotopic if there is a continuous family $(\varphi_t)_{t\in[0,1]}$ in $C^\infty(\mathbb{S}^{n-1})$, endowed with the C^∞ topology, which coincides with φ_0 at $t = 0$ and φ_1 at $t = 1$ such that $(1 + \Delta_{\Phi_t}^{(0)})^{-1}$ is norm continuous in $\mathcal{L}(L^2(\mathbb{R}^n))$, if Φ_t is associated with φ_t according to (11.29).

From the necessary condition of Proposition 11.11 iii) and since the space of compact operators $\mathcal{K}(L^2(\mathbb{R}^n))$ is closed in $\mathcal{L}(L^2(\mathbb{R}^n))$ we deduce the next result.

Proposition 11.18.
The set of Witten Laplacians $\Delta_\Phi^{(0)}$, $\Phi(r\theta) = r^m \varphi(\theta)$ for $r \geq 1$, $m > 1$ $\varphi \in C^\infty(\mathbb{S}^{n-1})$, has at least two homotopy classes.

Let $\mathcal{M}_m(\mathbb{S}^{n-1})$ denote the set of Morse functions on $\mathbb{S}^{n-1}$ for which 0 is not (the value of) a local minimum when $m > 2$ and the set of Morse functions for which 0 is not a critical value when $1 < m \leq 2$. It is an open dense set in $C^\infty(\mathbb{S}^{n-1})$. All its elements satisfy Condition (11.32). We call $\mathcal{M}_m$ the topological space of functions $\Phi \in C^\infty(\mathbb{R}^n)$ of the form $\Phi(r\theta) = r^m \varphi(\theta)$ for $r \geq 1$ with $\varphi \in \mathcal{M}_m(\mathbb{S}^{n-1})$, endowed with the $C^\infty(\{|x| \leq 1\})$-topology.

Lemma 11.19.
For $m > 1$ the application $\mathcal{M}_m \ni \varphi \to (1 + \Delta_\Phi^{(0)})^{-1} \in \mathcal{K}(L^2(\mathbb{R}^n))$, with $\Phi(r\theta) = r^m \varphi(\theta)$ $(r \geq 1)$ is norm continuous.

We can forget the behaviour of Φ in $B(0,C)$ because this leads to relatively compact perturbations. Since $(1 + \Delta_\Phi^{(0)})^{-1}$ is a non negative operator and $A \to A^\alpha$ is monotonous for $\alpha \in]0,1[$, it suffices to check that the map $\varphi \mapsto (1 + \Delta_\Phi^{(0)})^{-N}$ is continuous for $N \in \mathbb{N}$ large enough. By differentiating $(1 + \Delta_{\Phi_t}^{(0)})^{-N}$

with respect to $t \in [0,1]$ with $\varphi_t = (1-t)\varphi_0 + t\varphi_1$, we obtain a sum of terms of the form

$$(1+\Delta_{\Phi_t}^{(0)})^{-N_1} B(x)(1+\Delta_{\Phi_t}^{(0)})^{-N_2}, \qquad N_1 + N_2 = N+1, \qquad (11.36)$$

with $B \in C^\infty(\mathbb{R}^d)$ and $B(x) = O(\langle x \rangle^{2m-2})$. According to Remark 11.13, the operator $(1+\Delta_{\Phi}^{(0)})^{-1}$ is continuous from $L^{2,s}$ to $L^{2,s+\frac{m}{\mu}-1}$ with here $\mu = 1$ if $m \in]1,2]$ and $\mu \leq 2$ if $m > 2$. Moreover the constants in the semiclassical estimates (11.33)(11.34) can be taken uniform when one works in a neighborhood in the C^∞-topology of a given Morse function φ_0 (Note that here the semiclassical estimate (11.33) simply relies on some quadratic approximation technique and does not require the full hypoelliptic machinery). Thus the norm of $(1+\Delta_{\Phi}^{(0)})^{-1} : L^{2,s} \rightarrow L^{2,s+\frac{m}{\mu}-1}$ is uniformly bounded by some constant C_{φ_0} when φ belongs to some small neighborhood $\mathcal{N}_{\varphi_0}$ of a given Morse function φ_0. Hence the operator (11.36) is uniformly bounded for $N+1 \geq \frac{2m-2}{\frac{m}{\mu}-1}$ in such a neighborhood $\mathcal{N}_{\varphi_0}$. This implies the continuity of $(1+\Delta_{\Phi}^{(0)})^{-1}$ with respect to φ.

By considering the possible deformation of Morse functions while avoiding 0 as (the value of) a local minimum ($m > 2$) or a critical value ($1 < m \leq 2$), one proves the

Proposition 11.20.
For $m > 2$, the set of Witten Laplacian on 0-forms $\Delta_{\Phi}^{(0)}$, $\Phi \in \mathcal{M}_m$, has exactly two homotopy classes. For $m \in]1,2]$, there is an infinite number of homotopy classes.

In the case $m \in]1,2]$, the gradient of φ must not vanish on $\varphi^{-1}\{0\}$. Hence $\varphi^{-1}(]-\infty,0])$ is made of a finite number of connected C^∞ closed subset of $\mathbb{S}^{n-1}$. It is impossible to modify this number of connected components without having a vanishing gradient, hence without having a non compact resolvent for $\Delta_{\Phi}^{(0)}$. The number of connected components of $\varphi^{-1}(]-\infty,0)$ is an homotopy invariant in the case $m \in]1,2]$. In the case $m > 2$, one can reduce the number of connected components $\varphi^{-1}(]-\infty,0])$ to 1, if there is any, while staying in $\mathcal{M}_m$.

Semi-classical Analysis for the Schrödinger Operator: Harmonic Approximation

12.1 Introduction

The harmonic oscillator plays a crucial role in Quantum Mechanics. This is not only the fact that its spectrum can be computed explicitly. As we shall see here, it gives also the right approximation when analyzing the spectrum of the Schrödinger operator near the bottom in the generic situation where there is a unique non degenerate minimum. The harmonic approximation consists indeed in comparing in the one well case, the spectrum of $-h^2\Delta + V$ and the spectrum of the corresponding Harmonic oscillator obtained by replacing V by its quadratic approximation at the minimum. We shall analyze in this chapter this approximation, some large dimension aspects and also higher order approximations. We refer for this chapter to the books [CFKS], [Hel4] and [DiSj].

12.2 The Case of Dimension 1

We start with the simplest one-well problem:

$$S_v^h := -h^2 d^2/dx^2 + v(x) , \tag{12.1}$$

where v is a C^∞- function tending to ∞ and admitting a unique minimum at say 0 with $v(0) = 0$.
Let us assume that the minimum is non degenerate, i. e.

$$v''(0) > 0 . \tag{12.2}$$

In this very simple case, the justification of the harmonic approximation is an elementary exercise. We first consider the harmonic oscillator attached to 0:

$$-h^2 d^2/dx^2 + \frac{1}{2}v''(0)x^2 . \tag{12.3}$$

Using the dilation $x = h^{\frac{1}{2}} y$, we observe that this operator is unitarily equivalent to

$$h\left[-d^2/dy^2 + \frac{1}{2}v''(0)y^2\right] . \tag{12.4}$$

Consequently, the eigenvalues are given as

$$\lambda_n(h) = h \cdot \lambda_n(1) = (2n+1)h \cdot \sqrt{\frac{v''(0)}{2}} , \tag{12.5}$$

and the corresponding eigenfunctions are

$$u_n^h(x) = h^{-\frac{1}{4}} u_n^1\left(\frac{x}{h^{\frac{1}{2}}}\right) , \tag{12.6}$$

with [1]

$$u_n^1(y) = P_n(y) \exp -\sqrt{\frac{v''(0)}{2}} \frac{y^2}{2} . \tag{12.7}$$

We are just looking for simplification at the first eigenvalue. We consider the function $u_1^{h,app.}$

$$x \mapsto \chi(x)u_1^h(x) = c \cdot \chi(x) h^{-\frac{1}{4}} \exp -\sqrt{\frac{v''(0)}{2}} \frac{x^2}{2h} ,$$

where χ is compactly supported in a small neighborhood of 0 and equal to 1 in a smaller neighborhood of 0. The constant $c > 0$ is chosen such that $\|u_1^{h,app.} U\|_{L^2} = 1 + \mathcal{O}(h)$. We now get, the existence of C_0 and h_0 such that, for all $h \in (0, h_0)$,

$$\|(S_v^h - h \cdot \sqrt{\frac{v''(0)}{2}})u_1^{h,app.}\| \leq C_0\, h^{\frac{3}{2}} \|u_1^{h,app.}\| . \tag{12.8}$$

The coefficients corresponding to the commutation of S_v^h and χ give exponentially small terms. The main contribution in the computation of the L^2-norm of the l.h.s of (12.8) is

$$\|(v(x) - \frac{1}{2}v''(0)x^2)\chi(x)u_1^h(x)\|_{L^2} ,$$

which is easily seen as $\mathcal{O}(h^{\frac{3}{2}})$. Then the spectral theorem gives the existence for S_v^h of an eigenvalue $\lambda(h)$ such that

$$|\lambda(h) - h \cdot \sqrt{\frac{v''(0)}{2}}| \leq C_0 \cdot h^{\frac{3}{2}} ,$$

where C_0 was introduced in (12.8).

---

[1] We normalize by assuming that the L^2-norm is one. For the first eigenvalue, we have seen that, by assuming in addition that the function is strictly positive, we determine completely $u_1^h(x)$.

In particular, we get the inequality

$$\lambda_1(h) \leq h \cdot \sqrt{\frac{v''(0)}{2}} + Ch^{\frac{3}{2}} \,. \tag{12.9}$$

Remark 12.1.
At this level, we have only from (12.9) an inequality in one direction. If we analyze what we have done, we have used that $u_1^h(x)$ and its first derivative are exponentially small like $\mathcal{O}(\exp -\epsilon/h)$ in L^2 outside a neighborhood of 0 and the finer estimate $\|x^3 \cdot u_1^h(x)\|_{L^2} = \mathcal{O}(h^{\frac{3}{2}})$. If we want to exchange in the proof the roles of the harmonic oscillator and of S_v^h, we have to prove the same properties for the first eigenfunction of S_v^h without using the explicit expression of the first eigenfunction which is unknown for a general v. This will be done Chapter 13 through the proof of the so-called Agmon estimates.

Remark 12.2.
Note that the proof works also in the case of a potential with more than one minimum. Observe that with the introduction of χ, we have used only the behavior of v near one minimum.

Complete expansions.

We now sketch one way for getting a complete expansion for the eigenvalue (see also what we shall get later by WKB constructions). We will prove the following

Proposition 12.3.
Under assumptions (12.1)-(12.2), there exists, for any m, a normalized quasi-mode $u_1^{h,app.,m}$ and $\mu_1^{h,m}$ such that

$$(S_v^h - \mu_1^{h,m})u_1^{h,app.,m} = \mathcal{O}(h^{1+\frac{m+1}{2}}) \,.$$

We start with a completely formal expansion in powers of $h^{\frac{1}{2}}$ and x. For this we consider the Taylor expansion of v at 0:

$$v(x) \sim \sum_{k \geq 2} v_k x^k \,.$$

One will make more precise the feeling that everything depends only on this Taylor's expansion at the origin (modulo an error of order $\mathcal{O}(h^\infty)$).
Let us consider the formal operator

$$-h^2 d^2/dx^2 + \sum_{k \geq 2} v_k x^k \,.$$

Following what we have done for the harmonic approximation, we consider the dilation associated with the change of variables

$$x = h^{\frac{1}{2}}y \ .$$

In the new variables, we get the new formal operator

$$h\left(-d^2/dy^2 + \sum_{k\geq 2} h^{\frac{k}{2}-1}v_k y^k\right) \ ,$$

that we write in the form

$$h\left(\sum_{\ell\in\mathbb{N}} h^{\frac{\ell}{2}}T_\ell\right) \ ,$$

with

$$T_0 = -d^2/dy^2 + v_2 y^2 \ ,$$
$$T_\ell = v_\ell \, y^{\ell+2} \text{ for } \ell \geq 1 \ .$$

The formal problem consists in looking for a family $(w_\ell, \mu_{1,\ell})$ in $\mathcal{S}(\mathbb{R}) \times \mathbb{R}$ such that

$$\left(\sum_{\ell\in\mathbb{N}} h^{\frac{\ell}{2}}(T_\ell - \mu_{1,\ell})\right)\left(\sum_\ell h^{\frac{\ell}{2}}w_\ell\right) \sim 0 \ ,$$

when identifying the different powers of $h^{\frac{1}{2}}$.

The first equation is

$$(T_0 - \mu_{1,0})w_0 = 0 \ .$$

This is solved by taking as $\mu_{1,0}$ the first eigenvalue of the harmonic oscillator T_0 and as w_0 the first eigenfunction. Until now we have just reproduced the previous construction.

Let us now consider the second equation. We find

$$(T_0 - \mu_{1,0})w_1 = T_1 w_0 - \mu_{1,1}w_0 \ .$$

This new equation can only be satisfied if the right hand side is orthogonal to w_0. The range of $(T_0 - \mu_{1,0})$ is indeed the orthogonal of the kernel (according to the standard Fredhom theory). This condition gives

$$\mu_{1,1} = \langle T_1 w_0 \mid w_0 \rangle \ ,$$

that is, using the property that w_0 is even,

$$\mu_{1,1} = 0 \ .$$

Then we can solve the equation by adding the condition that w_1 is orthogonal to w_0. We write

$$w_1 = R'(\mu_{1,0})(T_1 w_0 - \mu_{1,1}w_0) = R'(\mu_{1,0})(T_1 w_0) \ ,$$

where

$$R'(\mu_{1,0}) = (T_0 - \mu_{1,0})^{-1} \ ,$$

on the orthogonal of $\mathbb{R} \cdot w_0$ and

$$R'(\mu_{1,0})w_0 = 0 \ .$$

w_1 is a priori[2] in B^2 but, using the "global ellipticity" of the harmonic oscillator (see [Hel0] or Section 4.6), it is clear that w_0 being in $\mathcal{S}(\mathbb{R}) = \cap_{k\in\mathbb{N}}B^k$, the same is true for w_1.

The next equations can be solved in the same way. The general equation is

$$(T_0 - \mu_{1,0})w_\ell = \sum_{k=0}^{\ell-1}(T_{\ell-k} - \mu_{1,\ell-k})w_k \ ,$$

and $\mu_{1,\ell}$ is obtained by

$$\mu_{1,\ell} = \sum_{k=1}^{\ell-1}\langle(T_{\ell-k} - \mu_{1,\ell-k})w_k \mid w_0\rangle + \langle T_\ell w_0 \mid w_0\rangle \ .$$

Remark 12.4.
If we observe that w_0 is even and that the operators T_ℓ conserve the parity for ℓ even and inverse the parity for ℓ odd, then we get that $\mu_{1,\ell} = 0$ for odd ℓ.

Remark 12.5.
If we have referred to functional analysis and to the regularity for the harmonic oscillator, it was just for presenting relatively general arguments working in a more general situation. In the particular case, we can work much more explicitly using the Hermite polynomials . If one for example looks at the equation corresponding to $\ell = 1$, one has just to express $y^3 w_0$ as a linear combination of the second eigenvector and of the fourth eigenvector of T_0 and we have an explicit expression for w_1 as a new combination of these two eigenvectors. It is then clear from this explicit expression that w_1 is in $\mathcal{S}(\mathbb{R})$ and actually of gaussian type.

From the formal construction to an approximate solution

For a given m, we now take for our approximate solution the candidate

$$u_1^{h,app.,m}(x) = h^{-\frac{1}{4}} \chi(x) \sum_{\ell=0}^{m} h^{\frac{\ell}{2}} w_\ell(x \cdot h^{-\frac{1}{2}}) \ , \tag{12.10}$$

---

[2] We recall that

$$B^k(\mathbb{R}^m) = \{u \in L^2(\mathbb{R}^m) \mid x^\alpha D_x^\beta u \in L^2 \ , \ \forall \alpha, \beta \in \mathbb{N}^m \text{ with } |\alpha| + |\beta| \leq k\} \ .$$

where χ is a C^∞ function with compact support and equal to 1 in a neighborhood of 0, and one gets easily the announced properties. This proves the proposition.

Decomposition by partition of unity:

The following decomposition is valid for any partition of unity:

$$\langle -\Delta u \mid u \rangle = \sum_j -\langle \Delta \phi_j u \mid \phi_j u \rangle - \sum_j \| \, |\nabla \phi_j| \, u \, \|^2 \qquad (12.11)$$

for all u in C_0^∞ if

$$\sum_j \phi_j^2 = 1 \, . \qquad (12.12)$$

This formula permits, modulo a remainder term, to control the lower bound of $-h^2\Delta + V$ by considering separately the Schrödinger operator in the support of a fixed ϕ_j. In our special case, the best is to use a h-dependent partition of unity constructed with (around the wells) functions of the form $\phi(x \cdot h^{-\frac{2}{5}})$. We shall explain this in detail for the case in large dimension in Section 12.4.

12.3 Quadratic Models

We look in this section at the quadratic case. The only point to observe is that, if A is a definite positive matrix on $\mathbb{R}^m$, then the spectrum of the Harmonic oscillator:

$$-\Delta + \langle Ax \mid x \rangle$$

can be explicitly computed.
One can indeed diagonalize A and if we denote by α_j the eigenvalues of A, we get that the harmonic oscillator is unitarily equivalent to:

$$-\Delta + \sum_j \alpha_j x_j^2 \, ,$$

and one is reduced to the one dimensional case. The spectrum is discrete and the eigenvalues are given by

$$\lambda_{p_1,\cdots,p_m} = \mathrm{Tr}\,\sqrt{A} + 2\sum_k p_k \sqrt{\alpha_k} \, .$$

Let us do explicit computations for the following model

$$y \mapsto Q(y) = \frac{b}{2} \sum_j y_j^2 + \frac{a}{2} \sum_{j \in \mathbb{Z}/m\mathbb{Z}} (y_j - y_{j+1})^2 \, . \qquad (12.13)$$

This potential is sometimes called the harmonic spin-chain potential. The corresponding matrix can be written as

$$A := \frac{1}{2}(b + 2a)Id - \frac{a}{2}(\tau_{+1} + \tau_{-1}) \, ,$$

where $(\tau_1 u)_k = u_{k-1}$ for $k = 1, \ldots, m$ (with the convention that $u_0 = u_m$). The spectrum of this matrix is well known. One way to prove that these eigenvalues are

$$\omega_i(m) = b + 2a \left(1 - \cos(2\pi \frac{(i-1)}{m}) \right) \, , \quad i = 1, \ldots, m \, ,$$

is to use the discrete Fourier transform defined on $\mathbb{C}^m$ by $a \mapsto \hat{a}$, with

$$\hat{a}(k) = \frac{1}{\sqrt{m}} \sum_{n=1}^{m} \exp(-2i\pi \frac{kn}{m}) a_n \, ,$$

in order to diagonalize A. The first eigenvalue of the corresponding harmonic oscillator is then given by

$$\lambda^{osc}(m) = \sum_{i=1}^{m} \sqrt{\frac{b}{2} + a \left(1 - \cos(2\pi \frac{(i-1)}{m}) \right)} \, . \tag{12.14}$$

The thermodynamic limit is then

$$\lim_{m \to \infty} \frac{\lambda^{osc}(m)}{m} = \int_0^1 \sqrt{\frac{b}{2} + a \left(1 - \cos(2\pi\theta) \right)} \, d\theta \, .$$

12.4 The Harmonic Approximation, Analysis in Large Dimension

We follow here the proof given in [CFKS] (initially given in [Sim2]). Additionally, we present the proof given in [Hel8] where the remainders are controlled on a model with respect to the dimension. This permits to control the thermodynamic limit.

If the upper bound is immediate by the use of quasimodes, the lower bound is usually more delicate. What is used here is an adapted partition of unity. Although this method is easier (and permits to follow more easily the dependence on other parameters), the proof of Helffer-Sjöstrand [HelSj7] gives better remainders.

We focus on the model

$$V^{(m)}(x) = \sum_{j=1}^{m} \left(v(x_j) + \frac{a}{2}|x_j - x_{j+1}|^2 \right) \, .$$

We first observe that

Lemma 12.6.
If v has a unique minimum on $\mathbb{R}$ at 0 (single well), the minimum of $V^{(m)}$ is unique and equal to $m \cdot \min v$.
If v is symmetric and has two minima on $\mathbb{R}$ at $\pm s$ (for some $s > 0$) (double well), then $V^{(m)}$ has two minima at the following two points:

$$x_{c\pm}(m) = \pm x_c(m) \quad \text{with} \quad x_c = s\,(1,1,1,....,1). \tag{12.15}$$

At x_c, the Hessian corresponds[3] to

$$Q_0(\bar{x}) = \frac{1}{2}v''(s)\sum_{k=1}^{m}\bar{x}_k^2 + \frac{a}{2}\sum_{k=1}^{m}(\bar{x}_k - \bar{x}_{k+1})^2 \tag{12.16}$$

where we have written:
$$x = \bar{x} + x_c \tag{12.17}$$

The harmonic approximation consists in replacing S_h by

$$H_h = -h^2\Delta_{\bar{x}} + Q_0(\bar{x}) .$$

We first look at the simple case when v has a unique minimum and

- $\min v = v(0) = 0$,
- $v''(x) \geq v''(0) > 0$,
  and
- $0 \leq v(x) \leq C \cdot \langle x \rangle^{N_1}$.

Proposition 12.7. *(the convex case)*
Under these assumptions, there exist positive constants C and h_0 such that, for all $h \in]0, h_0]$ and all $m \geq 1$, we have:

$$h\lambda^{app.}(m) \leq \lambda_1(m,h) \leq h\lambda^{app.}(m) + C \cdot m \cdot h^{\frac{3}{2}} , \tag{12.18}$$

where $\lambda^{app.}(m)$ is the first eigenvalue of the harmonic approximating oscillator

$$H_1 = -\Delta_{\bar{x}} + Q_0(\bar{x}) . \tag{12.19}$$

Remark 12.8.
The validity of the harmonic approximation has been already explained. What is analyzed in detail here is the simultaneous control of the remainder with respect to h and m.

We now look at the non convex case.

Proposition 12.9. *(the non convex case)*
Let us assume that v is a double well potential with two non degenerate minima at $\pm s$, with $v(\pm s) = 0$. Then there exist positive constants C and h_0 such that for all $h \in]0, h_0]$ and all $m \geq 1$, we have:

$$h\lambda^{app.}(m) - C \cdot m \cdot h^{\frac{6}{5}} \leq \lambda_1(m,h) \leq h\lambda^{app.}(m) + Cmh^{\frac{3}{2}} . \tag{12.20}$$

---

[3] we take $s = 0$ in the one well case.

Proof:

The lower bound on the first eigenvalue is no more evident because the lower bound by the quadratic approximation is no more true. The proof given by B. Simon [Sim2] does not give a satisfactory result and introduces a bad behavior of the remainder with respect to m at least if one takes the initial partition of the unity. The proof of the upper bound, on the contrary, is easier because we can work with explicit expressions. We refer to [HelSj7].

Proof of the lower bound:

The idea is to construct an adapted partition of unity. We treat here the case of the operator

$$S^{(m)}(h) = -h^2 \sum_{k=1}^{m} \partial_{x_k}^2 + \frac{a}{2} \sum_{k=1}^{m} (x_k - x_{k+1})^2 + \sum_{k=1}^{m} v(x_k) ,$$

(We recall the convention $x_{m+1} = x_1$).

The inequality we want to prove can be rewritten in the form

$$-C \cdot m \cdot h^{\frac{6}{5}} \le \lambda_1(m, h) - h \cdot \lambda^{app.}(m) , \qquad (12.21)$$

where $\lambda^{app.}(m)$ is the already computed first eigenvalue of

$$-\sum_{k=1}^{m} \partial_{x_k}^2 + \frac{a}{2} \sum_{k=1}^{m} (x_k - x_{k+1})^2 + \frac{v''(s)}{2} |x|^2 ,$$

which is directly obtained as $\mathrm{Tr}\, Q^{\frac{1}{2}}$ where $Q^{\frac{1}{2}}$ is the positive square root of the matrix Q attached to the quadratic form

$$\langle x|Q|x \rangle = \frac{a}{2} \sum_{k=1}^{m} (x_k - x_{k+1})^2 + \frac{v''(s)}{2} |x|^2 .$$

A partition of unity

Let θ be a function in $C_0^\infty(]-1, +1[)$ such that

$$0 \le \theta \le 1 \text{ with } \theta = 1 \text{ on }]-1/2, 1/2[.$$

We now introduce for $j \in \{-1, 1\}$ and $k \in \{1, ..., m\}$,

$$\phi_j(t, h) = \theta((t - js)/h^{\frac{2}{5}}) , \qquad (12.22)$$

and we choose θ and ϕ_0 such that the following relation is satisfied:

$$\sum_{j \in \{-1,0,1\}} \phi_j(t, h)^2 = 1 . \qquad (12.23)$$

We now associate to a multi-index α, with

$$\alpha = (\alpha_1, \alpha_2, \ldots, \alpha_m) \text{ with } \alpha_k \in \{-1, 0, 1\} \,,$$

the cut-off function:

$$\mathbb{R}^m \ni \mapsto \phi_\alpha(x, h) = \prod_{k=1}^{m} \phi_{\alpha_k}(x_k, h) \,.$$

We deduce from (12.22) and (12.23) the following properties:

$$\sum_\alpha \phi_\alpha(x, h)^2 = 1 \,, \tag{12.24}$$

and

$$\sum_\alpha \|\nabla \phi_\alpha(x, h)\|^2 \le C \cdot m \cdot h^{-\frac{4}{5}} \,, \tag{12.25}$$

where C is independent of h and m. The first step is a very elementary and old formula[4], which appears already in (12.11),

$$S(h) = \sum_\alpha \phi_\alpha S(h) \phi_\alpha - h^2 \sum_\alpha \|\nabla \phi_\alpha\|^2 \,. \tag{12.26}$$

According to (12.25), Formula (12.26) permits to reduce the proof of (12.21) to the determination of a constant C, such that, for all α, m, we have

$$-Cmh^{\frac{6}{5}} + h\,\lambda^{app.}(m) \le \inf_{\substack{u \in C_0^\infty(E_\alpha)\,, \\ \|u\| = 1}} \langle S(h)u \mid u \rangle_{L^2} \,, \tag{12.27}$$

where E_α is defined as

$$E_\alpha = \prod_{k=1}^{m} B_{\alpha_k} \,,$$

where,

- for $j = \pm 1$,

$$B_j = \{t \,, \ |t - j.s| \le h^{\frac{2}{5}}\} \,,$$

  and

- for $j = 0$, by

$$B_0 = \{t \,, \ |t - \ell.s| \ge \frac{1}{2} h^{\frac{2}{5}} \text{ for } \ell = \pm 1\} \,.$$

---

[4] This formula is called in the context of the analysis of the essential spectrum of many body Schrödinger operators, see for example the book by Cycon-Froese-Kirsch-Simon [CFKS], the "IMS localization formula".

We now observe that, if

$$|z - \ell s| \geq \frac{1}{2} h^{\frac{2}{5}} \text{ for } \ell = \pm 1 ,$$

there exists $\delta > 0$ s.t

$$v(z) \geq \delta h^{\frac{4}{5}} , \tag{12.28}$$

and we get consequently, for $x \in E_\alpha$, with $\alpha_k = 0$,

$$v(x_k) \geq \delta h^{\frac{4}{5}} . \tag{12.29}$$

We have now localized the problem in a box E_α. Of course all our constants have to be found independently of α, m and h. We now introduce, for a given α,

$$\mathcal{C}_\alpha = \{k \mid \alpha_k = 0\} . \tag{12.30}$$

Note that, when $\mathcal{C}_\alpha = \emptyset$, the box E_α is a product of intervals, each interval containing s or $-s$. In order to prove (12.27), we can, according to (12.29), bound $S(h)$ in E_α from below by $S'(h, \mathcal{C}_\alpha)$ where $S'(h, \mathcal{C})$ equals

$$S'(h, \mathcal{C}) = -h^2 \Delta + \frac{a}{2} \sum_{k=1}^{m} (x_k - x_{k+1})^2 + \sum_{k \notin \mathcal{C}} v(x_k) + \delta n_{\mathcal{C}} h^{\frac{4}{5}} , \tag{12.31}$$

and

$$n_{\mathcal{C}} = \text{Cardinal } \mathcal{C} . \tag{12.32}$$

We have now to prove

$$-C m h^{\frac{6}{5}} + h \lambda^{app.}(m) \leq \inf_{\substack{u \in C_0^\infty(E_\alpha), \\ \|u\| = 1}} \langle S'(h, \mathcal{C}_\alpha) u \mid u \rangle_{L^2} . \tag{12.33}$$

Now in each E_α, we can approach $v(x_k)$, for $k \notin \mathcal{C}_\alpha$, by its quadratic approximation $\frac{v''(s)}{2} . (x_k - \alpha_k s)^2$ with, for each term, an error of order $C (x_k - \alpha_k s)^3$, that is in E_α of order $C h^{\frac{6}{5}}$. But there are $(m - n_{\mathcal{C}_\alpha})$ terms of this type and this error is consequently bounded from above by $C m h^{\frac{6}{5}}$ in this quadratic approximation.

We are now reduced to the study of a family of harmonic oscillators. We introduce

$$S(h, \alpha) = -h^2 \Delta + \frac{a}{2} \sum_{k=1}^{m} (x_k - x_{k+1})^2 + \frac{v''(s)}{2} \sum_{k \notin \mathcal{C}_\alpha} (x_k - \alpha_k s)^2 + \delta n_{\mathcal{C}_\alpha} h^{\frac{4}{5}} ,$$

$$\tag{12.34}$$

and it is sufficient to prove the existence of h_0 such that for all $0 < h \leq h_0$, for all m and for all α s.t $\mathcal{C}_\alpha = \mathcal{C}$, we have

$$h.\lambda^{app.}(m) \leq \inf_{\left\{\begin{array}{c} u \in C_0^\infty(\mathbb{R}^m), \\ ||u|| = 1 \end{array}\right\}} \langle S(h,\alpha)u \mid u \rangle_{L^2} . \tag{12.35}$$

We observe that the left hand side corresponds by definition to the right hand side with $\mathcal{C} = \emptyset$ and $\{\alpha_k = 1 , \forall k\}$ or $\{\alpha_k = -1 , \forall k\}$.
Let us consider the operator $S(h, \alpha)$. The potential

$$V_\alpha(x) = \frac{a}{2} \sum_{k=1}^m (x_k - x_{k+1})^2 + \frac{v''(s)}{2} \sum_{k \notin \mathcal{C}_\alpha} (x_k - \alpha_k s)^2$$

is a polynomial of order less than two and admits, if $\mathcal{C}_\alpha$ is different of $\{1, ..., m\}$, a unique minimum $V_\alpha \geq 0$ and its quadratic part is (except through $\mathcal{C}_\alpha$) independent of α. If $\mathcal{C}_\alpha = \{1, ..., m\}$, the minimum is given by $x_k = x_{k+1}$, $k = 1, ..., m$.
After a translation, the operator $S(h, \alpha) - \delta\, n_{\mathcal{C}_\alpha}\, h^{\frac{4}{5}}$ becomes unitarily equivalent to:

$$-h^2 \Delta_y + \min V_\alpha + \frac{a}{2} \sum_{k=1}^m (y_k - y_{k+1})^2 + \frac{v''(s)}{2} \sum_{k \notin \mathcal{C}} y_k^2$$
$$\geq -h^2 \Delta_y + \frac{a}{2} \sum_{k=1}^m (y_k - y_{k+1})^2 + \frac{v''(s)}{2} \sum_{k \notin \mathcal{C}} y_k^2 ,$$

with $\mathcal{C} = \mathcal{C}_\alpha$.
We are now reduced to a problem which depends only on α through $\mathcal{C} = \mathcal{C}_\alpha$. The last step is in the proof of the following

Lemma 12.10.
There exists h_0, such that, for all m and all $\mathcal{C}$, for all $u \in C_0^\infty(\mathbb{R}^m)$, we have

$$\langle \left(-h^2 \Delta_y + \frac{a}{2} \sum_{k=1}^m (y_k - y_{k+1})^2 + \frac{v''(s)}{2} \sum_{k \notin \mathcal{C}} y_k^2 + \delta\, n_{\mathcal{C}}\, h^{\frac{4}{5}}\right) u \mid u \rangle$$
$$\geq h\, \lambda^{app.}(m) ||u||^2 . \tag{12.36}$$

Proof:
Because we know explicitly the spectrum of the harmonic oscillator, we have to prove:

$$h\, \mathrm{Tr}\, (Q_{\mathcal{C}}^{\frac{1}{2}}) + \delta\, n_{\mathcal{C}}\, h^{\frac{4}{5}} \geq h\, \mathrm{Tr}\, (Q^{\frac{1}{2}}) , \tag{12.37}$$

where $Q_{\mathcal{C}}$ is the matrix attached to the quadratic form

$$Q_{\mathcal{C}}(y) = \frac{a}{2} \sum_{k=1}^m (y_k - y_{k+1})^2 + \frac{v''(s)}{2} \sum_{k \notin \mathcal{C}} y_k^2 , \quad Q = Q_\emptyset .$$

The problem is of course to control the uniformity with respect to m. We shall now prove the existence of C_0, such that, for all m and $\mathcal{C}$

$$\mathrm{Tr}\, (Q_{\mathcal{C}}^{\frac{1}{2}}) + n_{\mathcal{C}}\, C_0 \geq \mathrm{Tr}\, (Q^{\frac{1}{2}}) , \tag{12.38}$$

It is then not difficult to deduce that (12.33) is satisfied for $h \leq (\frac{\delta}{C_0})^5$.
It is a direct consequence of the following algebraic lemma.

Lemma 12.11.
Let A and B be two positive $p \times p$ matrices (symmetric real matrices or self-adjoint complex matrices) such that, for some $c > 0$,

- $A \geq c \operatorname{Id}$,
- $A - B \geq 0$.

Then

$$\operatorname{Tr}\left[A^{1/2} - B^{1/2}\right] \leq \frac{1}{\sigma} \|A - B\|_1 \ ,$$

where σ is the smallest eigenvalue of $A^{1/2} + B^{1/2}$ and $\|C\|_1 = \operatorname{Tr}\left(\sqrt{C^*C}\right)$.

Proof:
We will use a commutator argument. We write

$$A^{1/2} - B^{1/2} = \left(A - B + \left[A^{1/2}, B^{1/2}\right]\right)\left(A^{1/2} + B^{1/2}\right)^{-1} ,$$

and

$$A^{1/2} - B^{1/2} = \left(A^{1/2} + B^{1/2}\right)^{-1}\left(A - B - \left[A^{1/2}, B^{1/2}\right]\right) .$$

We take the trace of the sum and use the cyclicity of the trace. We get immediately

$$\operatorname{Tr}\left(A^{1/2} - B^{1/2}\right) = \operatorname{Tr}\left(\left(A^{1/2} + B^{1/2}\right)^{-1}(A - B)\right) ,$$

which yields

$$\operatorname{Tr}\left(A^{1/2} - B^{1/2}\right) \leq \left\|\left(A^{1/2} + B^{1/2}\right)^{-1}\right\| \operatorname{Tr}(A - B) \leq \frac{1}{\sigma}\operatorname{Tr}(A - B) .$$

∎

We apply the lemma with $A = Q$ and $B = Q_C$. We compute

$$\|A - B\|_1 = \frac{1}{2}v''(s)n_C$$

and we can choose

$$\sigma = \left(\frac{v''(s)}{2}\right)^{1/2} \quad \text{and} \quad C_0 = \frac{v''(s)}{(2v''(s))^{1/2}} = (\tfrac{1}{2}v''(s))^{1/2}.$$

Remark 12.12.
The analysis of the excited states is much more difficult and will strongly depend on the properties of v. Except in the case of a strictly uniformly convex case where the splitting between the second eigenvalue and the first one has a uniform lower bound [Sj3, HelSj7], general fine results with explicit control with respect to the dimension are still missing (see however [Hel9, Hel10]).

Let us also mention the large dimension analysis of excited states of Matte-Moeller [MaMo], where the authors play with the structure of the Witten Laplacians and the so-called supersymmetric argument (see Subsection 11.2.2).

13

Decay of Eigenfunctions
and Application to the Splitting

13.1 Introduction

As we have already seen when comparing the spectra of the harmonic oscillator
and of the Schrödinger operator, it could be quite important to know **a priori**
how the eigenfunction of the Schrödinger operator, associated to an eigenvalue
$\lambda(h)$, decays in the classically forbidden region, i.e. outside the classical region
$V^{-1}(]-\infty, \lambda(h)])$ as $h \to 0$.
The Agmon [Ag] estimates give a very efficient[1] way to control such a decay.
We refer to [Hel7] or to the original papers [Sj2, Sj3] and [HelSj7] for details
and complements on this point. Here we will recall a more standard material
that one can find in the original paper [HelSj1] or in the books [Hel4], [DiSj].

13.2 Energy Inequalities

The main but basic tool is a very simple identity attached to the Schrödinger
operator

$$S_h = -h^2 \Delta + V .$$

Proposition 13.1.
*Let Ω be a bounded open domain in $\mathbb{R}^m$ with C^2 boundary. Let $V \in C^0(\overline{\Omega}; \mathbb{R})$
and ϕ a real valued lipschitzian function on $\overline{\Omega}$. Then, for any $u \in C^2(\overline{\Omega}; \mathbb{R})$
with $u_{/\partial\Omega} = 0$, we have*

$$h^2 \int_\Omega |\nabla(\exp \tfrac{\phi}{h} u)|^2 \, dx + \int_\Omega (V - |\nabla\phi|^2) \exp 2\tfrac{\phi}{h} u^2 \, dx = \int_\Omega \exp 2\tfrac{\phi}{h} (S_h u)(x) \cdot u(x) \, dx . \tag{13.1}$$

Proof:
In the case when ϕ is a $C^2(\overline{\Omega})$- function, this is an immediate consequence of
the Green-Riemann formula

---

[1] cf the contributions of Helffer-Sjöstrand [HelSj1] and Simon [Sim2] in the 80's.

$$\int_\Omega |\nabla v|^2 \, dx = - \int_\Omega \Delta v \cdot v \, dx \,, \tag{13.2}$$

for all $v \in C^2(\overline{\Omega})$ such that $v_{/\partial\Omega} = 0$.

This can actually be extended to $u \in H_0^1(\Omega)$. To treat the general case, we just write ϕ as a limit as $\epsilon \to 0$ of $\phi_\epsilon = \chi_\epsilon \star \phi$ where $\chi_\epsilon(x) = \epsilon^{-m} \chi(\frac{x}{\epsilon})$ is the standard mollifier and we remark that, by Rademacher's Theorem, $\nabla\phi$ is almost everywhere the limit of $\nabla\phi_\epsilon = \nabla\chi_\epsilon \star \phi$.

13.3 The Agmon Distance

The Agmon metric attached to an energy E and a potential V is defined as $(V - E)_+ dx^2$ where dx^2 is the standard[2] metric on $\mathbb{R}^m$. This metric is degenerate. It is indeeed identically 0 at points living in the classical region:

$$U_E := \{x \mid V(x) \le E\} \,. \tag{13.3}$$

Associated to the Agmon metric, we define a natural distance

$$(x, y) \mapsto d_{(V-E)_+}(x, y)$$

by taking the infimum of the length of piecewise C^1 paths connecting x and y:

$$d_{(V-E)_+}(x, y) = \inf_{\left\{\begin{array}{c} \gamma \\ \gamma(0) = x ,\ \gamma(1) = y \end{array}\right\}} \int_0^1 [(V(\gamma(t)) - E)_+]^{\frac{1}{2}} \, |\gamma'(t)| \, dt \,. \tag{13.4}$$

When there is no ambiguity, we shall write more simply $d_{(V-E)_+} = d$. Similarly to the Euclidean case, we obtain the following properties:

- Triangular inequality

$$|d(x', y) - d(x, y)| \le d(x', x) \,, \ \forall x, x', y \in \mathbb{R}^m \,; \tag{13.5}$$

-

$$|\nabla_x d(x, y)|^2 \le (V - E)_+(x) \text{ a.e. } x \,. \tag{13.6}$$

We observe that the second inequality is satisfied for other distances like

$$d(x, U) = \inf_{y \in U} d(x, y) \,.$$

The most useful case will be the case when, for a given energy E, U equals U_E.

---

[2] One has also a natural extension on a complete Riemannian manifold.

13.4 Decay of Eigenfunctions for the Schrödinger Operator

When u_h is a normalized eigenfunction of the Dirichlet realization in Ω satisfying

$$S_h u_h = \lambda_h u_h ,$$

then (13.1) gives roughly that the function $\exp \frac{\phi}{h} \, u_h$ is well controlled (in $L^2(\Omega)$) in a region

$$\Omega_1(\epsilon_1, h) = \{ x \mid V(x) - |\nabla \phi(x)|^2 - \lambda_h > \epsilon_1 > 0 \} ,$$

by $\exp \left(\sup_{\Omega \setminus \Omega_1(\epsilon_1, h)} \frac{\phi(x)}{h} \right)$. The choice of a suitable ϕ (possibly depending on h) is related to the Agmon metric $(V - E)_+ \cdot dx^2$, when $\lambda_h \to E$ as $h \to 0$. Of course it could seem more natural to consider $(V - \lambda_h)_+ dx^2$ for optimal results but we prefer to work with a fix energy E. The typical choice is $\phi(x) = (1 - \epsilon)d(x)$, where $d(x)$ is the Agmon distance to U_E:

$$d(x) := d_{(V-E)_+}(x, U_E) . \tag{13.7}$$

In this case we get that the eigenfunction is localized inside a small neighborhood of U_E and we can measure the decay of the eigenfunction outside U_E by

$$\exp(1 - \epsilon)\frac{d(x)}{h} \, u_h = \mathcal{O}(\exp \frac{\epsilon}{h}) , \tag{13.8}$$

for any $\epsilon > 0$.
More precisely we get for example the following theorem

Theorem 13.2.
Assume that V is C^∞, semibounded and satisfies

$$\lim_{|x| \to \infty} \inf V > \inf V = 0 , \tag{13.9}$$

and

$$V(x) > 0 \text{ for } |x| \neq 0 . \tag{13.10}$$

Let u_h be a (family of) normalized eigenfunctions such that

$$S_h u_h = \lambda_h u_h , \tag{13.11}$$

with $\lambda_h \to 0$ as $h \to 0$. Then for all ϵ and all compact $K \subset \mathbb{R}^m$, there exists a constant $C_{\epsilon,K}$ such that, for h small enough,

$$\|\nabla(\exp \frac{d}{h} \cdot u_h)\|_{L^2(K)} + \| \exp \frac{d}{h} \cdot u_h\|_{L^2(K)} \leq C_{\epsilon,K} \exp \frac{\epsilon}{h} , \tag{13.12}$$

where $x \to d(x)$ is the Agmon distance between x and 0 attached to the Agmon metric $V \cdot dx^2$.

Useful improvements in the case when $E = \min V$ and when the minima are non degenerate can be obtained by controlling more carefully with respect to h. It leads for example in (13.12) to an upperbound in $\mathcal{O}(h^{-N})$ instead of $C_{\epsilon,K} \exp \frac{\epsilon}{h}$ in a neighborhood of the minimum. It is also possible to control the eigenfunction at ∞. This was actually the initial goal of S. Agmon [Ag].

Proof:

Let us choose some $\epsilon > 0$. We shall use the identity (13.1) with

- V replaced by $V - \lambda_h$,
- $\phi = (1 - \delta)\, d(x, U_E)$, with δ small enough depending on ϵ,
- $u = u_h$,
- $S_h = -h^2 \Delta + V - \lambda_h$.

Let

$$\Omega_\delta^+ = \{x \in \Omega \,,\, V(x) \geq \delta\}$$

and

$$\Omega_\delta^- = \{x \in \Omega \,,\, V(x) < \delta\}\,.$$

We deduce from (13.1)

$$h^2 \int_\Omega |\nabla(\exp \tfrac{\phi}{h} u_h)|^2 dx + \int_{\Omega_\delta^+} (V - \lambda_h - |\nabla\phi|^2) \exp \tfrac{2\phi}{h} u_h^2 \, dx$$
$$\leq \sup_{x \in \Omega_\delta^-} |V(x) - \lambda_h - |\nabla\phi|^2| \left(\int_{\Omega_\delta^-} \exp \tfrac{2\phi}{h} u_h^2 \, dx \right)\,.$$

Then, for some constant C independent of $h \in]0, h_0]$ and $\delta \in]0,1]$, we get

$$h^2 \int_\Omega |\nabla(\exp \tfrac{\phi}{h} u_h)|^2 dx + \int_{\Omega_\delta^+} (V - \lambda_h - |\nabla\phi|^2) \exp \tfrac{2\phi}{h} u_h^2 \, dx$$
$$\leq C \cdot \left(\int_{\Omega_\delta^-} \exp \tfrac{2\phi}{h} u_h^2 \, dx \right)\,.$$

Let us observe now that on Ω_δ^+ we have (with $\phi = (1-\delta)d(\cdot, U)$):

$$V - \lambda_h - |\nabla\phi|^2 \geq (2 - \delta)\delta^2 + o(1)\,.$$

Choosing $h(\delta)$ small enough, we then get for any $h \in]0, h(\delta)]$

$$V - \lambda_h - |\nabla\phi|^2 \geq \delta^2\,.$$

This permits to get the estimate

$$h^2 \int_\Omega |\nabla(\exp \tfrac{\phi}{h} u_h)|^2 dx + \delta^2 \int_{\Omega_\delta^+} \exp \tfrac{2\phi}{h} u_h^2 \, dx \leq C \cdot \left(\int_{\Omega_\delta^-} \exp \tfrac{2\phi}{h} u_h^2 \, dx \right)\,,$$

and finally

$$h^2 \int_\Omega |\nabla(\exp \tfrac{\phi}{h} u_h)|^2 dx + \delta^2 \int_\Omega \exp \tfrac{2\phi}{h} u_h^2 \, dx \leq \tilde{C} \cdot \exp \tfrac{a(\delta)}{h}\,,$$

where $a(\delta) = 2 \sup_{x \in \Omega_\delta^-} \phi(x)$. We now observe that $\lim_{\delta \to 0} a(\delta) = 0$ and the end of the proof is then easy.

Application: Comparison of two Dirichlet problems

Let us consider two open sets Ω_1 and Ω_2 containing a unique well[3] $U_E = V^{-1}(] - \infty, E])$ attached to an energy E. If for example $\Omega_1 \subset \Omega_2$, the Agmon estimates permit to prove the existence of a bijection b between the spectrum of $S_{(h, \Omega_1)}$ in an interval $I(h)$ tending (as $h \to 0$) to E and the corresponding spectrum of $S_{(h, \Omega_2)}$ such that $|b(\lambda) - \lambda| = \mathcal{O}(\exp - \frac{S}{h})$ (under a weak assumption on the spectrum at $\partial I(h)$). S is here any constant such that

$$0 < S < d_{(V-E)_+}(\partial \Omega_1, U) .$$

One can simply use that any eigenfunction of one operator can be used, after multiplication by a cut-off function, as an approximate eigenfunction for the other one, and the a priori Agmon decay estimates permits to control the error due to the cut-off.

This can actually be improved (using more sophisticated perturbation theory) as $\mathcal{O}(\exp - \frac{2S}{h})$.

13.5 Estimates on the Resolvent

Although the proof of the decay estimates was done differently in [HelSj1] (See [Hel4]), the systematic use of resolvent estimates was decisive for all the other papers [HelSj2, HelSj3, HelSj4]. The use of the Agmon estimates for $(S_h - z)$, where z avoids the spectrum, leads indeed to useful estimates on the resolvent (see [HelSj2] or Proposition 6.5 in [DiSj]) that we shall briefly describe. We consider for simplicity the case when V has a unique well U (of diameter 0 for the Agmon distance associated to the energy 0). Let us first give some definitions.

Definition 13.3.
Let $\mathcal{J} \subset]0, 1]$ such that $0 \in \overline{\mathcal{J}}$ and let $A = \mathbf{A_h}(h \in \mathcal{J})$ a family of bounded operators from $L^2(M)$ into $H^1(M)$. Here M is a C^∞-riemannian manifold (possibly with boundary). Let $f \in C^0(M \times M ; \mathbb{R})$. We say that the family of distribution kernels $A_h(x, y; h)$ associated[4] to $\mathbf{A_h}$ is $\tilde{O}(\exp - \frac{f(x,y)}{h})$, if the following property is satisfied:
$\forall x_0, y_0 \in M, \epsilon > 0$, there exist neighborhoods $V \subset M$ of x_0, $U \subset M$ of y_0 and $C_\epsilon > 0$ s.t.:

$$||\mathbf{A_h} u||_{H^1(V)} \leq C_\epsilon \exp - \frac{[f(x_0, y_0) - \epsilon]}{h} ||u||_{L^2(U)} ,$$

---

[3] This means that U_E is connected
[4] We recall that this means that $\langle \mathbf{A_h} \phi \mid \psi \rangle_{L^2} = A_h(\phi \otimes \psi)$, where $(\phi \otimes \psi)(x, y) = \phi(x)\psi(y)$.

for all $u \in L^2(M)$ such that supp $u \subset U$.

As a consequence, we get the continuity of the operator $\mathbf{A_h}$ in weighted spaces with control of the norms with respect to h. Typically, if φ and ψ are continuous functions on M such that:

$$-\varphi(x) \leq f(x,y) - \psi(y) ,$$

then, for any $\epsilon > 0$, there exists C_ϵ and h_ϵ such that, for $h \in]0, h_\epsilon]$,

$$\|\mathbf{A_h} u\|_{L^2(M, \exp -\frac{2\varphi}{h} dx)} \leq C_\epsilon \exp \frac{\epsilon}{h} \|u\|_{L^2(M, \exp -\frac{2\psi}{h} dx)} , \quad \forall u \in C^\infty(M) .$$

We can now state the main result:

Proposition 13.4.
Under the above assumptions, we have, for h small enough,

$$(S_h - z)^{-1}(x, y) = \tilde{O}(\exp -\frac{d(x,y)}{h}) , \tag{13.13}$$

uniformly, with respect to z satisfying the conditions

$$z \in B(V(U), \epsilon_h) , \quad \epsilon_h \to 0 ,$$

and

$$d(z, \sigma(S_h)) \geq \frac{1}{C_\epsilon} \exp -\frac{\epsilon}{h} , \quad \forall \epsilon > 0 .$$

We will give an application of this proposition in Section 13.8.

13.6 WKB Constructions

Although the harmonic approximation (and its refinement given in Proposition 12.3) is satisfactory for determining the asymptotics (modulo $\mathcal{O}(h^\infty)$) of a fixed number of low lying eigenvalues, it does not give the exact behavior of the corresponding eigenfunctions. This can be immediately seen in the 1-dimensional case. In the analysis of finer effects like the splitting between the two lowest eigenvalues (in the case for multiple wells problems), we need a better information on the eigenfunctions of some reference problems "far" from the well which in suitable coordinates is the point 0. This is done by the construction of WKB solutions. We will remain rather sketchy and refer to [DiSj] (or to the original paper [HelSj1] for details). These solutions are obtained in the form:

$$u^{wkb}(x, h) = h^{-\frac{n}{4}} a(x, h) \exp -\frac{\phi(x)}{h} , \tag{13.14}$$

where ϕ is a real phase such that

$$\phi \geq 0 \, , \, \phi(0) = 0 \, , \tag{13.15}$$

and where the amplitude $a(x, h)$ admits an expansion:

$$a(x, h) \sim \sum_{j=0}^{+\infty} a_j(x) h^j \, , \tag{13.16}$$

where we take as initial condition:

$$a_0(0) \neq 0 \, , \, a_j(0) = 0 \, , \, \forall j > 0 \, . \tag{13.17}$$

The condition on ϕ is natural: we would like to have an approximate eigenfunction which is well localized at 0 and the eigenfunction in the case of the harmonic oscillator has this form with $\phi(x) = \alpha x^2$ ($\alpha > 0$).
One is now looking for a formal eigenvalue

$$E(h) \sim \sum_{j=0}^{+\infty} E_j h^j \, , \tag{13.18}$$

with $E_0 = \inf V$, and such that formally, that is at the level of the formal series expansion in powers of h,

$$\exp \frac{\phi(x)}{h} (S_h - E(h)) u^{wkb}(x, h) \sim 0 \, . \tag{13.19}$$

The identity,

$$e^{\frac{\phi(x)}{h}} (h^2 \Delta) e^{-\frac{\phi(x)}{h}} = h^2 \Delta + h \left(-2\nabla\phi.\nabla - \Delta\phi \right) + |\nabla\phi|^2 \, ,$$

leads first (by expressing the cancellation of the coefficient of h^0)

$$|\nabla\phi(x)|^2 = V(x) - \inf V \, , \tag{13.20}$$

and to, once this first equation, called the eikonal equation, is satisfied,

$$-h^2 \Delta a + 2h(\nabla\phi \cdot \nabla a + \Delta\phi\, a) - \left(\sum_{j>0} E_j \right) a \sim 0 \, . \tag{13.21}$$

The second one will be cancelled term by term (according to the powers of h) and will lead to the so called transport equations. It was proved in [HelSj1] (see also [DiSj]) that the eikonal equation (13.20), with the additional condition (13.15), admits a $\mathcal{C}^\infty$ solution near the bottom of V when the minimum is non degenenerate. Fix this non degenerate minimum at $x = 0$. The function ϕ is defined as the generating function of the Lagrangian manifold in $T^*\mathbb{R}^m$, denoted by Λ, of which the existence is given by the stable manifold theorem. This manifold is defined in a neighborhood of $(0,0)$ as

$$\Lambda = \{(x, \xi) \in T^*\mathbb{R}^m \mid \lim_{t \to -\infty} \Phi_t(x, \xi) = (0,0)\} \, ,$$

where Φ_t is the flow of the hamiltonian H_q associated to $q(x,\xi) = -\xi^2 + V(x)$. By construction this manifold is invariant by H_q. To say that ϕ is a generating function means that locally

$$\Lambda = \Lambda_\phi = \{(x, \nabla\phi(x)) \mid x \in \mathcal{V}(0)\} \ .$$

One notices that the integral curves of H_q in $q^{-1}(0)$ are projected by the map $(x,\xi) \mapsto x$ on the integral curves of $\nabla\phi$. One can also verify that

$$\phi(x) = d_{(V-\inf V)}(x, 0) \tag{13.22}$$

and satisfies

$$\phi(x) \sim \frac{1}{2}\left(x \cdot \operatorname{Hess}(V(0))^{\frac{1}{2}}x\right) + \mathcal{O}(|x|^3) \ ,$$

and

$$\nabla\phi(x) \sim \operatorname{Hess}(V(0))^{\frac{1}{2}}x + \mathcal{O}(|x|^2) \ ,$$

as $x \to 0$.

Thus, one can take ϕ as a C^∞ function in any open set Ω such that each point of Ω is connected to 0 by a unique minimal geodesic contained in Ω. We will call these domains **geodesically-starshape domains (with respect to 0)**.

Let us consider now the amplitude. Once ϕ is known, the equation (13.21) leads after term by term identification to the inductive system of transport equations, for $j \geq 0$,

$$\nabla\phi(x) \cdot \nabla a_j + (\Delta\phi(x) - E_1)a_j = \Delta a_{j-1} + \sum_{k+\ell=j+1,\ k\geq 2} E_k a_\ell \ , \tag{13.23}$$

where we set $a_{-1} = 0$.

Here are two problems to solve:

1) The constants E_j have to be determined for $j \geq 1$;

2) The vector field $\nabla\phi$ vanishes and the transport equations cannot be solved by simply dressing[5] the vector field.

The proof will be in two steps.

First one can solve recursively these equations modulo flat functions at the level of formal series. This will lead to the determination of the E_j's and of the Taylor expansions of the a_j's, taking $a_0(0) = 1$, $a_j(0) = 0$ for $j \geq 1$ as initial conditions. Taking for example the first equation, which reads:

$$\nabla\phi(x) \cdot \nabla a_0 + (\Delta\phi(x) - E_1)a_0 = 0 \ , \tag{13.24}$$

we get as a necessary condition $\Delta\phi(0) = E_1$, which is of course compatible with what we found by the harmonic approximation.

---

[5] i.e. by finding a change of variable such that in the new cordinates the vector field becomes simply $\frac{d}{dx_1}$,

Secondly we can solve the transport equation by integration along the integral curves $g_t(x)$ $(t \in]-\infty, 0])$ of $\nabla\phi$ having 0 as limiting point as $t \to -\infty$ on flat functions. We refer the reader to [HelSj1], [Hel4] or [DiSj] for details.

The transport equations are then solved recursively in the same way. We observe that like for the solution of the eikonal equation, we can also solve the transport equations, not only in a small neighborhood of 0, but also in any neighborhood of 0 which is geodesically-starshape with respect to 0.

Once we have a formal series a_j, we can take a realization $a(x, h)$ by a Borel procedure, which leads to the construction of a WKB solution. In suitable open sets, one can first get through the spectral theorem that this WKB solution gives (after renormalization) in geodesically-starshape open sets a very good approximation of the one-well ground state:

$$\|u_h^1 - u_h^{wkb}\|_{H^1(\Omega)} = \mathcal{O}(h^\infty) \,.$$

Using simple Agmon's type estimates[6] , one can show that these estimates propagate along the integral curves of $\nabla\phi(x)$ which leads to

$$\| \exp \frac{\phi}{h}(u_h^1 - u_h^{wkb})\|_{H^1(\Omega)} = \mathcal{O}(h^\infty) \,. \tag{13.25}$$

Again we refer the reader to [HelSj1], [Hel4] or [DiSj] for the details.

13.7 Upper Bounds for the Splitting Between the Two First Eigenvalues

The control of the decay of the eigenfunctions has also immediate consequences on the splitting of eigenvalues. We shall first give a rough estimate, with a proof which only works for the splitting between the two first eigenvalues of S_h. A more powerful technique will be presented in Section 13.8.

13.7.1 Rough Estimates

To understand what is needed, let us recall the following classical formula for the splitting which is nothing else than a version of the minimax principle applied to the orthogonal of $\mathbb{R} \cdot u_1$,

$$\lambda_2 - \lambda_1 =$$

$$\inf_{\left\{ \begin{array}{c} \phi \,;\, \phi \in C_0^\infty, \\ \int \phi(x)\, u_1(x)^2\, dx = 0 \end{array} \right\}} \left[\left(\int |h\nabla\phi|^2\, u_1(x)^2\, dx \right) / \left(\int |\phi|^2\, u_1(x)^2\, dx \right) \right] \,. \tag{13.26}$$

---

[6] For the specialist in microlocal analysis let us mention that these estimates are strongly related to the proof of microlocal propagation of regularity for microhyperbolic operators [Sj1] and [Mar].

Here u_1 denotes the first normalized eigenfunction of the Schrödinger operator. The estimates about the splitting are then deduced from a good choice of ϕ and from a precise information on the decay of u_1 in suitable domains.

Let us now consider the double well situation. This means that the potential v is symmetric:
$$v(-x) = v(x) \ ,$$
and has two non degenerate minima $\pm x_c$. We now choose a function ϕ in C_0^∞ which satisfies
$$\phi(x) = -\phi(-x)$$
and
$$\phi = 1$$
in a neighborhood of the critical point x_c of V. We recall also that the first eigenfunction is even
$$u_1(x) = u_1(-x) \ .$$
We have used here that the first eigenvalue is simple and that the first eigenfunction can be chosen strictly positive (Perron-Frobenius or Krein-Rutman argument).

Hence we have
$$\int \phi(x) u_1(x)^2 dx = 0 \ .$$

We observe also using the Agmon estimates (13.12) that u_1 is exponentially small (as $h \to 0$) on the support of the functions $(1 - \phi^2)$ and $|\nabla \phi|$, so
$$\int \phi(x)^2 u_1(x)^2 \ dx = 1 + \mathcal{O}(\exp -\frac{S}{h}) \ ,$$
and that
$$\int |\nabla \phi(x)|^2 u_1(x)^2 dx = \mathcal{O}(\exp -\frac{S}{h}) \ ,$$
for some $S > 0$.

Coming back to the formula giving the splitting we obtain
$$\lambda_2 - \lambda_1 = \mathcal{O}(\exp -\frac{S}{h}) \ . \tag{13.27}$$

In the situation of a symmetric double well, we have consequently two eigenvalues whose difference is exponentially small.

Remark 13.5.
Considering more carefully the result concerning the decay of the first eigenfunction (see (13.12)), one can choose any S such that
$$0 < S < d_{(V-E)_+}(-x_c, x_c) := S_{-+} \ . \tag{13.28}$$

13.7.2 Towards More Precise Estimates

In some generic cases, one can actually give a more precise estimate for the splitting in the form

$$\lambda_2 - \lambda_1 = h^{\frac{1}{2}} A(h)(\exp -\frac{S_{-+}}{h}) , \tag{13.29}$$

where S_{-+} is the Agmon distance between the two wells and $A(h)$ is a non zero function admitting an expansion of the type

$$A(h) \sim \sum_{j=0}^{+\infty} a_j h^j ,$$

with $a_0 \neq 0$.

This will be explained in Section 13.8 through the computation of the so-called interaction matrix.

13.7.3 Historical Remarks

The asymptotics (13.29) for the splitting was obtained in this form in [HelSj1] and in a weaker form but sooner by Jona-Lasinio, Martinelli, and Scoppola [JoMaSc81] and also by B. Simon [Sim2]. Independently a rather sketchy proof was given by V.P. Maslov in 1984 [Mas84]. Actually there is a long history on the subject. The rigorous study of the splitting, which was present very soon after the beginning of the foundations of the Quantum Mechanics (see Landau-Lifschitz) was motivated, as mentioned for example in the notes of Reed-Simon [ReSi], by the problems posed by M. Kac [Kac]. In the case $m = 1$, the first rigorous proof is given by E. Harrell in 1980 [Har], using the theory of ordinary differential equations and by Combes-Duclos-Seiler [CoDuSe] at the same time.

We just mention the "instantons" method as for example presented by Coleman [Cole] which did not lead to completely rigorous proofs untill recently.

We emphasize that the "difficult" part in the study of the splitting is to get a lower bound.

For the specific operators that we have in mind in these notes, there is also a lot of results obtained by probabilistic methods (see [HolKusStr], [BovEckGayKl1, BovEckGayKl2] and [BovGayKl]). We will come back to this point in Chapter 15.

13.8 Interaction Matrix
for the Symmetic Double Well Problem

Once the harmonic approximation is done, it is possible to construct an orthonormal basis of the spectral space attached to a given interval $I(h) :=$

[inf V, inf $V + Ch$] (C avoiding some discrete values), each of the elements of the basis being exponentially localized in one of the wells.
The computation of the matrix of the operator in this basis, using WKB approximations, leads to the so-called "interaction matrix" (See Dimassi-Sjöstrand [DiSj] or Helffer [Hel4]).

We consider the case with two wells, say U_1 and U_2. We assume that there is a symmetry[7] g in $\mathbb{R}^m$, such that $g^2 = Id$, $gU_1 = U_2$, and such that the corresponding action on $L^2(\mathbb{R}^m)$ defined by $gu(x) = u(g^{-1}x)$ commutes with the Laplacian. In addition $gV = V$.
We now define reference one well problems by introducing:

$$M_1 = \mathbb{R}^m \setminus B(U_2, \eta) , \; M_2 = \mathbb{R}^m \setminus B(U_1, \eta) .$$

With this choice, we have $gM_1 = M_2$. The parameter $\eta > 0$ is free but can always be chosen arbitrarily small. We denote by ϕ_j the corresponding ground state of the Dirichlet realization of $-h^2 \Delta + V$ in M_j and corresponding to the ground state energy $\lambda_{M_1} = \lambda_{M_2}$. According to our result on the decay recalled in (13.12), these eigenfunctions decay like $\tilde{O}(\exp - \frac{d(x, U_j)}{h})$, where $\tilde{O}(f)$ roughly[8] means $\exp \frac{\epsilon}{h} \cdot \mathcal{O}_\epsilon(f)$ for all $\epsilon > 0$ as $h \to 0$. We can of course keep the relation

$$g\phi_1 = \phi_2 .$$

Let us now introduce θ_j, which is equal to 1 on $B(U_j, \frac{3}{2}\eta)$ and with support in $B(U_j, 2\eta)$. We introduce

$$\chi_1 = 1 - \theta_2 , \; \chi_2 = 1 - \theta_1 ,$$

and we can also keep the symmetry condition:

$$g\chi_1 = \chi_2 .$$

Our approximate eigenspace will be generated by

$$\psi_j = \chi_j \phi_j , \; (j = 1, 2) ,$$

which satisfies

$$S_h \psi_j = \lambda_M \psi_j + r_j ,$$

with

$$r_j = h^2 (\Delta \chi_j) \phi_j + 2h^2 (\nabla \chi_j) \cdot (\nabla \phi_j) .$$

We note that the "smallness" of r_j can be immediately controlled using the decay estimates (13.12) in $B(U_j, 2\eta) \setminus B(U_j, \frac{3}{2}\eta)$.

In order to construct an orthonormal basis of the eigenspace F corresponding to the two lowest eigenvalues near λ_M, we first project our basis ψ_j which was not far to be orthogonal and introduce:

---

[7] Typically, we take $g = -Id$.

[8] More precisely, for any $\epsilon > 0$, one can choose above $\eta > 0$ such that...

$$v_j = \Pi_F \psi_j .$$

The resolvent formula shows that $v_j - \psi_j$ can be made very small (at least $\exp -\frac{S}{h}$ with S satisfying (13.28) by chosing $\eta > 0$ small enough). More precisely, we have the following comparison.

Lemma 13.6.

$$(v_j - \psi_j)(x) = \tilde{O}(\exp -\frac{\delta_j(x)}{h}) , \qquad (13.30)$$

in $\mathbb{R}^m \setminus B(U_{\hat{\jmath}}, 4\eta)$, where $\hat{1} = 2$, $\hat{2} = 1$ and

$$\delta_j(x) = d(x, U_{\hat{\jmath}}) + d(U_1, U_2) .$$

Proof
Our starting point is:

$$S_{h, M_j} \psi_j = \lambda_{M_j} \psi_j + r_j .$$

where

$$\operatorname{supp} r_j \subset B(U_{\hat{\jmath}}, 2\eta) ,$$

and

$$r_j = \tilde{O}(\exp -\frac{d(x, U_j)}{h}) .$$

We have $v_j - \Pi_F \psi_j \in F^\perp$ and the spectral theorem gives already the estimate

$$\|v_j - \pi_F \psi_j\| = \tilde{O}(\exp -\frac{d(U_1, U_2)}{h}) .$$

For a suitable contour Γ_h in $\mathbb{C}$ containing the interval $I(h)$ and remaining at a suitable distance of the spectrum

$$d(\Gamma_h, \sigma(S_h)) \geq \frac{1}{C_\epsilon} \exp -\frac{\epsilon}{h}, \forall \epsilon > 0 , \qquad (13.31)$$

we can write:

$$v_j - \psi_j = \frac{1}{2\pi} \int_{\Gamma_h} (\lambda_M - z)^{-1} (S_h - z)^{-1} r_j dz .$$

We observe by referring to (13.13) that:

$$(S_h - z)^{-1} r_j = \tilde{O}(\sup_{y \in \operatorname{supp} r_j} \exp -\frac{[d(x,y) + d(y, U_j)]}{h})$$
$$= \tilde{O}(\exp -\frac{\delta_j(x)}{h}) .$$

The separation assumption (13.31) permits to get the same property for $v_j - \psi_j$.

Remark 13.7.
We notice that:

$$\delta_j(x) \geq d(x, U_j) \,,$$

What we see here is that the improved estimate does not lead to improvements near $U_{\hat{j}}$, where we have modified ϕ_j into ψ_j by introducing a cut-off function but that the improvement is quite significative when keeping a large distance (compare to η) with $U_{\hat{j}}$.

It is sometimes useful (for the treatment[9] of "resonant wells"), to relax the assumption by introducing the weaker property that there exists $a > 0$, such that, for all $\epsilon > 0$, there exists $C_\epsilon > 0$, such that:

$$d(\Gamma_h, \sigma(S_h)) \geq \frac{1}{C_\epsilon} \exp -\frac{a - \epsilon}{h} \,. \tag{13.32}$$

In this case one has still an estimate by replacing δ_j by $\delta_j - a$.

We then orthonormalize by the Gram-Schmidt procedure.

$$e_j = \sum_k (V^{-\frac{1}{2}})_{jk} v_k \,,$$

with

$$V_{ij} = \langle v_i \mid v_j \rangle \,.$$

We note that

$$V_{ij} - \delta_{ij} = \mathcal{O}(\exp -\frac{S}{h}) \,.$$

At each step, we control the difference $e_j - \psi_j$, which satisfies also (13.30). The matrix we would like to analyze is then simply the two by two matrix

$$M_{ij} = \langle (S_h - \lambda_M)e_i \mid e_j \rangle \,.$$

The eigenvalues of this matrix measure the dispersion of the two eigenvalues around λ_M.

We observe that symmetry considerations lead to:

$$M_{12} = M_{21} \text{ and } M_{11} = M_{22} \,.$$

So the eigenvalues are easy to compute and corresponding eigenvectors are $\frac{1}{\sqrt{2}}(1,1)$ and $\frac{1}{\sqrt{2}}(1,1)(-1,+1)$. As soon as we have the main behavior of M_{12}, we can deduce that the eigenvalues are simple and that the splitting between the two eigenvalues is given by $2|M_{12}|$.

---

[9] A non-resonant well U_k ($k \neq j$) is a well for which the renormalized (after division by h) lowest eigenvalue of the harmonic approximation at U_k is strictly above the corresponding lowest eigenvalue at U_j.

It remains to explain how one can compute M_{12}. The analysis of the decay permits to show that

$$M_{12} = \frac{1}{2} \left(\langle r_2 , \psi_1 \rangle + \langle r_1 , \psi_2 \rangle \right) + \mathcal{R}_{12} , \tag{13.33}$$

with

$$\mathcal{R}_{12} = \mathcal{O}(\exp -\frac{2S}{h}) , \tag{13.34}$$

for a suitable choice of $\eta > 0$ small enough.

An integration by parts leads (observing that $\nabla\chi_1 \cdot \nabla\chi_2 \equiv 0$ for our choice of η) to the formula

$$M_{12} = h^2 \int \chi_1 (\phi_2 \nabla\phi_1 - \phi_1 \nabla\phi_2) \nabla\chi_2 + \mathcal{R}_{12} . \tag{13.35}$$

A priori informations on the decay permit to restrict the integration in the right hand side of (13.35) to the set $\{d(x, U_1) + d(x, U_2) \le d(U_1, U_2) + a\}$ for some $a > 0$.

A computation based on the Stokes Lemma gives then the existence of $\epsilon_0 > 0$ such that:

$$M_{12} = h^2 \int_\Gamma [\phi_2 \partial_n \phi_1 - \phi_1 \partial_n \phi_2] d\nu_\Gamma + \mathcal{O}(\exp -\frac{S_{12} + \epsilon_0}{h}) . \tag{13.36}$$

Here $S_{12} = d(U_1, U_2)$ and Γ is an open piece of hypersurface defined in the neighborhood of the minimal geodesic $\mathrm{geod}(U_1, U_2)$ between the two points U_1 and U_2, that we assume for simplification to be unique and ∂_n denotes the normal derivative to Γ, positively oriented from U_1 to U_2 .

The last step is to observe that in a neighborhhood of the intersection γ_{12} of Γ with $\mathrm{geod}(U_1, U_2)$, one can replace the function ϕ_j (or ψ_j) modulo $\mathcal{O}(h^\infty) \exp -\frac{d(x, U_j)}{h}$ by its WKB approximation $h^{-\frac{n}{4}} a_j(x, h) \exp -\frac{d(x, U_j)}{h}$.

This leads finally to

$$\begin{aligned} M_{12} = &\; h^{1-\frac{n}{2}} \exp -\frac{d(U_1, U_2)}{h} \times \\ &\times \int_\Gamma \exp -\frac{(d(x, U_1) + d(x, U_2) - d(U_1, U_2))}{h} \times \\ &\times (a_1(x, 0) a_2(x, 0) (\partial_n d(x, U_1) - (\partial_n d(x, U_2)) + \mathcal{O}(h)) \, d\nu_\Gamma , \end{aligned} \tag{13.37}$$

where $d\nu_\Gamma$ is the induced measure on Γ.

With natural generic additional assumptions saying that the map

$$\Gamma \ni x \mapsto (d(x, U_1) + d(x, U_2) - d(U_1, U_2))$$

vanishes exactly at order 2 at γ_{12}, this finally leads to Formula (13.29), after use of the Laplace integral method.

Semi-classical Analysis and Witten Laplacians: Morse Inequalities

The aim of this chapter is to see how the technique of the harmonic approximation permits to analyze roughly the smallest eigenvalues of the Witten Laplacians. This will permit us to split between the eigenvalues of the Witten Laplacians which are $o(h)$ – and which will appear to be actually $\mathcal{O}(h^{\frac{3}{2}})$– and the others for which we will show that they do not belong to an interval $(-\infty, \epsilon_0 h]$ for some $\epsilon_0 > 0$.

This is an important step towards a finer analysis of exponentially small effects and has already nice applications to the Morse theory that we recall briefly in order to fix the framework. The material presented here is now standard. We refer to the initial paper by E. Witten [Wi], the book by Cycon-Froese-Kirsch-Simon [CFKS], the original article of Helffer-Sjöstrand [HelSj4], the book by Helffer [Hel4] and the recent book by W. Zhang [ZH] which develops more the topological aspects.

14.1 De Rham Complex

Let M be a compact C^∞ Riemannian oriented n-dimensional manifold and let f be a C^∞ map from M into $\mathbb{R}$. We shall say that f is a Morse function if all its critical point are non degenerate. Let d be the differential on M and let $\Lambda^p(M)$ denote the space of the C^∞ p-forms. We have a natural scalar product $\Lambda^p(M)$ and can take its completion for the associate norm in order to get an Hilbert space $\Lambda^p L^2(M)$ of L^2-sections. The restriction of d to Λ^p is denoted by $d^{(p)}$:

$$\Lambda^p(M) \ni \omega \mapsto d^{(p)}\omega \in \Lambda^{p+1}(M) \,, \tag{14.1}$$

and we denote by $d^{(p)^*}$ the formal adjoint sending $\Lambda^{p+1}(M)$ into $\Lambda^p(M)$. d^* is the differential operator on

$$\Lambda(M) = \oplus_{p=0}^n \Lambda^p(M) \,, \tag{14.2}$$

whose restriction to $\Lambda^p(M)$ is $d^{(p-1)^*}$.

We observe that

$$d^2 = 0 \, , \tag{14.3}$$

which expresses the property that d is a complex (called the de Rham complex).

It is possible to show that for any p, $\mathrm{Im}\, d^{(p-1)}$ is a subspace of $\mathrm{Ker}\, d^{(p)}$, with finite codimension. This leads to the definition of the Betti numbers:

Definition 14.1.
For any p, we define b_p as the codimension of $\mathrm{Im}\, d^{(p-1)}$ in $\mathrm{Ker}\, d^{(p)}$.

We can then define the de Rham Laplacian by

$$\Delta^{DR} = (d + d^*)^2 = dd^* + d^*d \, , \tag{14.4}$$

By restriction to the C^∞ p-forms, we get the Laplace-Beltrami operator on the p-forms:

$$\Delta^{(p)} = d^{(p)^*} d^{(p)} + d^{(p-1)} d^{(p-1)^*} \, . \tag{14.5}$$

This operator is an elliptic operator of order 2 and when M is compact is essentially self-adjoint on $\Lambda^p(M)$. The resolvent is compact and in particular the kernel of $\Delta^{(p)}$ has a finite dimension and one can show that the p-th Betti number satisfies:

$$b_p = \mathrm{dim}\ \mathrm{Ker}\,(\Delta^{(p)}) \, . \tag{14.6}$$

The classical Hodge theory gives a natural bijection between $\mathrm{Ker}\, d^{(p)} / \mathrm{Im}\, d^{(p-1)}$ and $\mathrm{Ker}\,\Delta^{(p)}$.

If $\omega \in \mathrm{Ker}\, d^{(p)}$, we write $\omega = \sigma_1 + \sigma_1^\perp$ with $\sigma_1 \in \mathrm{Ker}\,\Delta^{(p)}$ and $\sigma_1^\perp$ is given by

$$\sigma_1^\perp = d^{(p-1)} \sigma_2 \, ,$$

with

$$\sigma_2 = d^{(p-1)^*} (\Delta^{(p)})^{-1,'} \omega \, .$$

Here $(\Delta^{(p)})^{-1,'}$ is defined as being equal to 0 on $\mathrm{Ker}\,\Delta^{(p)}$ and to $(\Delta^{(p)})^{-1}$ on $(\mathrm{Ker}\,\Delta^{(p)})^\perp$.

14.2 Useful Formulas

We recall some formulas which will be useful for introducing our Witten complexes and the corresponding Laplacians. We follow the appendix of [HelSj4] (See Arnold [Arn] Chapter 7) but everything is very standard. The reader can also look in [CFKS] for a rather selfcontained presentation.

For a function u and a 1-form ω, we have:

$$du = \sum_{j=1}^n (\partial_{x_j} u) dx_j \, ,$$

$$d^* \omega(x_0) = - \sum_{j=1}^n \partial_{x_j} \omega_j(x_0) \quad \text{(in normal coordinates around } x_0\text{)}.$$

Then we can extend the definition to any form in order that the following formula is satisfied, for any p-form ω and any q-form θ:

$$d(\omega \wedge \theta) = d\omega \wedge \theta + (-1)^p \omega \wedge d\theta \ .$$

In particular, we have

$$
\begin{aligned}
d(\textstyle\sum_j a_j dx^j) &= \textstyle\sum_j (da_j) \wedge dx_j \\
&= \textstyle\sum_{j,k} (\partial_{x_k} a_j) dx_k \wedge dx_j \\
&= \textstyle\sum_{j<k} (\partial_{x_j} a_k - \partial_{x_k} a_j) dx_j \wedge dx_k \ .
\end{aligned}
$$

When considering the Witten complex, we start from:

$$d_f = d + df \wedge \ ,$$

and its formal Hilbertian adjoint:

$$d_f^* = d^* + \nabla f \lrcorner$$

For a vector field X, the operator $\omega \mapsto X \lrcorner \omega$ is also denoted by i_X. It sends the ℓ-forms into the $(\ell - 1)$-forms. In particular we have

$$i_X(\omega) = <\omega, X> \ , \quad i_X(du) = (Xu) \ .$$

Note that, if ω is a p-form and θ is a q-form, then

$$i_X(\omega \wedge \theta) = (i_X \omega) \wedge \theta + (-1)^p \omega \wedge (i_X \theta) \ .$$

A basic formula in differential geometry is:

$$\mathcal{L}_X = i_X \, d + d \, i_X \ , \tag{14.7}$$

where $\mathcal{L}_X$ is the Lie derivative along X.

On functions, we recall that we have simply $\mathcal{L}_X \, u = Xu$ (the second term vanishes). In general, the Lie derivative is usually defined by differentiation along the flow Φ_t of X, by

$$\mathcal{L}\omega = \frac{d}{dt}(\Phi_t^* \omega)_{/t=0} \ ,$$

but in our context we can take the right hand side of (14.7) as a definition. We observe that

$$d \, \mathcal{L}_X = \mathcal{L}_X \, d = d \, i_X \, d \ .$$

Note that

$$\mathcal{L}_X(\omega \wedge \theta) = (\mathcal{L}_X \omega) \wedge \theta + \omega \wedge (\mathcal{L}_X \theta) \ .$$

14.3 Computation of the Witten Laplacian on Functions and 1-Forms

Let us show how the above calculus permit to give the expression of the Witten Laplacian[1]. We have

$$\Delta_f = (d_f + d_f^*)^2$$
$$= d_f d_f^* + d_f^* d_f$$
$$= dd^* + d^* d + \nabla f \lrcorner\, df \wedge + df \wedge \nabla f \lrcorner + \mathcal{L}_{\nabla f} + \mathcal{L}_{\nabla f}^* \,.$$

Computation of M_f.

Let us compute $M_f := \mathcal{L}_{\nabla f} + \mathcal{L}_{\nabla f}^*$ in simple cases.

On functions:

$$\mathcal{L}_{\nabla f} g = \nabla f \cdot \nabla g \,.$$

The adjoint of $\mathcal{L}_{\nabla f}$ is easily calculated by

$$\mathcal{L}_{\nabla f}^* g = -\nabla f \cdot \nabla g - (\Delta f)\, g \,.$$

So

$$M_f g = -(\Delta f)\, g \,.$$

This last operator is a multiplication operator by $-\Delta f$.

On 1-forms (in $\mathbb{R}^n$):

$$\mathcal{L}_{\nabla f}\Big(\sum_{j=1}^{n} a_j dx_j\Big) = \sum_{j=1}^{n} (\mathcal{L}_{\nabla f} a_j) dx_j + \sum_{j=1}^{n} a_j (\mathcal{L}_{\nabla f} dx_j) \,.$$

So we have just to compute $(L_{\nabla f} dx_j)$ using the fundamental formula.

$$\mathcal{L}_X dx_j = dX_j = \sum_{k=1}^{n} (\partial_k X_j)\, dx_k \,, \qquad (14.8)$$

that we apply with $X = \nabla f$. This gives, for a 1-form $\Phi\, d\mu$,

$$\mathcal{L}_{\nabla f}(\Phi d\mu) = (\nabla f \nabla \Phi)\, d\mu \;+\; \Phi \operatorname{Hess} f d\mu \,.$$

One can then compute the adjoint. We have to compute:

$$\langle \mathcal{L}_X^*(g d\lambda) \mid (\phi d\mu)\rangle = \langle (g d\lambda) \mid \mathcal{L}_X(\phi d\mu)\rangle$$

We can then apply the previous formula.

---

[1] For conciseness, these computations are presented in the case of $\mathbb{R}^n$ with the euclidean metric. See the final remark for the more general case and [CFKS] for details.

$$\langle \mathcal{L}_X^*(g d\lambda) \mid (\phi d\mu) \rangle = < g \mid X\phi >_{L^2} \langle d\lambda \mid d\mu \rangle + \langle (g d\lambda) \mid \phi \mathcal{L}_X(d\mu) \rangle$$

We first observe that, if $X = \nabla f$,

$$\mathcal{L}_X^*(g d\lambda) = (\mathcal{L}_X^* g) d\lambda + g \operatorname{Hess} f d\lambda .$$

This leads to:

$$\mathcal{L}_{\nabla f}^*(g d\lambda) = (-\nabla f \nabla g) d\lambda + g \left(\operatorname{Hess} f d\lambda - \Delta f d\lambda \right) .$$

Then we find:

$$(\mathcal{L}_{\nabla f} + \mathcal{L}_{\nabla f}^*) = -(\Delta f) + 2(\operatorname{Hess} f) .$$

The other computation is that, on 1-forms:

$$\nabla f \lrcorner \, df \wedge + df \wedge \nabla f \, \lrcorner = |\nabla f|^2$$

(because it is a linear local operator, it is enough to verify the formula for $\omega = dx_j$).

So we get on 1-forms

$$\begin{aligned}
\Delta_f^{(1)} &= dd^* + d^* d + |\nabla f|^2 \otimes I + 2\operatorname{Hess} f - \Delta f \otimes I \\
&= \Delta^{(1)} + |\nabla f|^2 \otimes I + 2\operatorname{Hess} f - \Delta f \otimes I \\
&= \Delta^{(0)} \otimes I + |\nabla f|^2 \otimes I + 2\operatorname{Hess} f - \Delta f \otimes I ,
\end{aligned}$$

and finally:

$$\Delta_f^{(1)} = \Delta_f^{(0)} \otimes I + 2 \operatorname{Hess} f . \tag{14.9}$$

Remark 14.2.
The computation is more complicated on a manifold !! The Laplace-Beltrami operator $\Delta^{(1)}$ is no more equal to $\Delta^{(0)} \otimes I$ but to the sum $\mathcal{B}^{(1)} + \mathcal{R}_{(4)}$, where $\mathcal{B}^{(1)}$ is the Bochner Laplacian on 1-forms and $\mathcal{R}_{(4)}$ is the Ricci tensor. This formula is valid for p-forms, $\Delta^{(p)} = \mathcal{B}^{(p)} + \mathcal{R}$. In normal coordinates around a point x_0, this leads to the formula

$$\Delta^{(p)} = \Delta^{(0)} \otimes \operatorname{Id} + A_{(2)}(x, \partial_x) + \mathcal{R}_{(4)}(x) , \tag{14.10}$$

where $A_{(2)}$ is a first order differential operator with vanishing principal part at x_0. We refer the reader for example to [CFKS]-Theorem 12.20 for the details.

14.4 The Morse Inequalities

When f is a Morse function on a compact manifold M, that is if all the critical points of f are non degenerate, we can associate to each critical point U_j the index ℓ_j corresponding to the number of $(-)$ in the signature of the Hessian of f at U_j. For example, the index is 0 if f has a non degenerate minimum at

U_j and n if f has a non degenerate maximum at U_j.
For $\ell = 0, \ldots, n$, we denote by:

$$C^{(\ell)} = \{j \mid U_j \text{ is of index } \ell\} . \tag{14.11}$$

In this case, we shall sometimes write $U_j^{(\ell)}$ in order to indicate the reference to the index. We can then define:

$$m_\ell = \#C^{(\ell)} . \tag{14.12}$$

The so called weak Morse inequalities say:

Theorem 14.3.
Let f be a Morse function on a compact manifold M. Then with the above notation (14.12) we have

$$m_\ell \geq b_\ell , \quad \text{for } \ell = 0, \ldots, n . \tag{14.13}$$

Another (more precise) way to formulate these inequalities is to say that there exists a complex $\overset{\circ}{E}$ of finite dimensional vector spaces:

$$0 \to E^0 \overset{u_0}{\to} E^1 \to \cdots E^k \overset{u_k}{\to} E^{k+1} \to \cdots \overset{u_{n-1}}{\to} E^n \to 0 , \tag{14.14}$$

with

$$\dim_{\mathbb{R}} E^k = m_k$$

and

$$\dim_{\mathbb{R}} H^k(E) := \dim (\operatorname{Ker} u_k / \operatorname{Im} u_{k-1}) = b_k .$$

This leads by elementary algebra to the Strong Morse Inequalities:

Theorem 14.4.
Under the assumptions of Theorem 14.3, we have, for any i such that $0 \leq i \leq n$,

$$b_i - b_{i-1} + \cdots + (-1)^i b_0 \leq m_i - m_{i-1} + \cdots + (-1)^i m_0 . \tag{14.15}$$

Moreover we have:

$$b_n - b_{n-1} + \cdots + (-1)^n b_0 = m_n - m_{n-1} + \cdots + (-1)^i m_0 . \tag{14.16}$$

We refer the reader [Mil] or [Lau] for topological proofs of the Morse inequalities.

Remark 14.5.
When $M = \mathbb{R}^n$, one can have a similar result under the conditions that $|\nabla f| \geq \frac{1}{C}$ outside a ball of radius C and $-\Delta f \geq -C|\nabla f|^2$.

14.5 The Witten Complex

Following Witten [Wi], we would like to find such a complex of finite dimensional vector spaces having the same cohomology as the de Rham complex. For this purpose, we introduce a perturbation of the de Rham complex now called the Witten complex depending on a parameter $h > 0$:

$$d_{f,h} = \exp -\frac{f}{h}(hd) \exp \frac{f}{h} . \qquad (14.17)$$

It is clear that the new complex has the same Betti numbers as the de Rham complex:

$$\dim \left(\operatorname{Ker} d_{f,h}^{(p)} / \operatorname{Im} d_{f,h}^{(p-1)} \right) = b_p . \qquad (14.18)$$

The Hodge theory works in this new context. In particular, one can introduce

$$\Delta_{f,h} = (d_{f,h} + d_{f,h}^*)^2 , \qquad (14.19)$$

which gives by restriction on the p-forms $\Delta_{f,h}^{(p)}$. For each $h > 0$ and any C^∞ function f, $\Delta_{f,h}^{(p)}$ has the same properties as $\Delta^{(p)}$.

In particular, for an interval $I(h)$, let us consider $E_f^{(p)}(I(h))$ the eigenspace of $\Delta_{f,h}^{(p)}$ associated to the eigenvalues belonging to the interval $I(h)$.

From the Hodge theory we get:

$$b_p = \dim \operatorname{Ker} \Delta_{f,h}^{(p)} , \qquad (14.20)$$

and, if $I(h) \ni 0$, we have the trivial inclusion:

$$\operatorname{Ker} \Delta_{f,h}^{(p)} \subset E_f^{(p)}(I(h)) . \qquad (14.21)$$

We first take $A > 0$ and

$$I_A(h) = [0, Ah^{\frac{3}{2}}] . \qquad (14.22)$$

As an immediate consequence, we get

$$b_p \leq \dim E_f^{(p)}(I_A(h)) , \ \forall A > 0 , \ \forall h > 0 . \qquad (14.23)$$

The proof of the weak Morse inequalities is now reduced to the proof that there exist $A > 0$ and $h > 0$ such that:

$$\dim E_f^{(p)}(I_A(h)) = m_p . \qquad (14.24)$$

The semi-classical analysis will permit to show that,

Proposition 14.6.
For any $A > 0$, there exists $h_0(A)$ such that $\forall h \in]0, h_0(A)]$ such that (14.24) is satisfied.

Let us also observe (see (11.11)) that we have:

$$d_{f,h} \circ \Delta_{f,h} = \Delta_{f,h} \circ d_{f,h} \ . \tag{14.25}$$

This implies that for any interval $I(h)$:

$$d_{f,h}^{(p)} E_f^{(p)}(I(h)) \subset E_f^{(p+1)}(I(h)) \ , \tag{14.26}$$

and by taking the adjoint

$$d_{f,h}^{(p),*} E_f^{(p+1)}(I(h)) \subset E_f^{(p)}(I(h)) \ , \tag{14.27}$$

Then $\overset{\circ}{d_{f,h}}$ reduced to the complex $\overset{\circ}{E_f}(I(h))$ defines a complex of finite dimensional spaces whose Betti numbers are the same as for the de Rham complex. As observed above this gives the strong Morse Inequalities if we can perform the semi-classical analysis.

14.6 Rough Semi-classical Analysis of the Witten Laplacian

As we have explained in the previous section, the proof of the Morse inequalities is based on rough estimates on the bottom of the spectrum of the Witten Laplacians modulo $\mathcal{O}(h^{\frac{3}{2}})$. According to the semi-classical theory presented in Chapter 12, we just have to use the harmonic approximation at each critical point of f. We observe indeed that on $\mathbb{R}^n$ with the euclidean metric each Witten Laplacian $\Delta_{f,h}^{(p)}$ has the form

$$\Delta_{f,h}^{(p)} = \Delta_{f,h}^{(0)} \otimes I + h V_{1,f,p} \ , \tag{14.28}$$

where $x \mapsto V_{1,f,p}(x)$ is matrix-valued. Here we have:

$$\Delta_{f,h}^{(0)} = -h^2 \Delta + |\nabla f|^2 - h \Delta f \ . \tag{14.29}$$

The Witten Laplacian on C^∞ functions has the structure of a Schrödinger operator, where the electric potential V_h has the form:

$$V_h(x) = V_0(x) + h V_1(x) \ , \tag{14.30}$$

where the main term:

$$V_0(x) = |\nabla f(x)|^2 \ , \tag{14.31}$$

admits as minima the critical points of f.

The Witten Laplacians on p-forms have the same structure. They can more precisely be written as:

$$\Delta_{f,h}^{(p)} = -h^2 \Delta^{(p)} + |\nabla f|^2 + h(\mathcal{L}_{\nabla f} + \mathcal{L}_{\nabla f}^*) \,, \tag{14.32}$$

where the Laplace-Beltrami operator $\Delta^{(p)}$ on p-forms equals $\mathcal{B}^{(p)} + \mathcal{R}_{(4)}$ according to Remark 14.2 and $\mathcal{L}_{\nabla f}$ is the Lie derivative along ∇f.

Remark 14.7.
The terms $h^2 A_{(2)}(x, \partial_x)$ and $h^2 \mathcal{R}_{(4)}(x)$ coming from (14.10) only bring higher order corrections to the harmonic approximation (see [CFKS] or [HelSj4]).

Harmonic approximation for the Witten Laplacian

We have already computed $M_f := (\mathcal{L}_{\nabla f} + \mathcal{L}_{\nabla f}^*)$ for $p = 0$ and 1. More generally, the computation of the matrix M_f at a critical point U_j of f gives:

$$M_f(U_j) = 2 \, d\Gamma^p (\mathrm{Hess}\, f(U_j)) - \mathrm{Tr}\, (\mathrm{Hess}\, f(U_j)) \,, \tag{14.33}$$

where $d\Gamma^p(A)$ is the natural action[2] of a symmetric matrix $A \in \mathcal{L}(E)$ on a p-form in $\Lambda^p E$.

Remark 14.8.
In the case of $M = \mathbb{R}^n$, formula (14.33) is true at any point.

Note in particular that if $\lambda_1, \cdots, \lambda_n$ are the eigenvalues of a symmetric matrix B (with $\lambda_1 \leq \lambda_2 \leq \cdots \leq \lambda_n$) then the eigenvalues of $d\Gamma^p(B)$ are of the form $\lambda_{i_1} + \cdots + \lambda_{i_p}$ with $i_1 < i_2 < \cdots < i_p$.

The ground state energy of the Harmonic oscillator approximating $\Delta_{f,h}^{(p)}$ at the critical point U_j is given by:

$$\mu_1 := \sum_{j=1}^n |\lambda_j| + 2(\lambda_1 + \lambda_2 + \cdots + \lambda_p) - \sum_{j=1}^n \lambda_j \,,$$

where the λ_j's $(j = 1, \ldots, n)$ denote the eigenvalues of $\mathrm{Hess}\, f(U_j)$, ordered in increasing order:
$$\lambda_1 \leq \lambda_2 \leq \cdots \leq \lambda_n \,.$$

Let us analyze when this quantity vanishes.
When $p = 0$, it is clear that it vanishes if and only if all the eigenvalues are strictly positive. This corresponds to the points of index 0. As a consequence, the dimension of the eigenspace corresponding to the small eigenvalues of $\Delta_{f,h}^{(0)}$ (i.e. for eigenvalues lying in $[0, \epsilon_0 h]$ for some sufficiently small $\epsilon_0 > 0$ or for

---

[2] This natural action is defined by recursion:

$$d\Gamma^{(1)}(A) = A \text{ and } d\Gamma^{(p)}(A)(u_1 \wedge u^{(p-1)}) = A u_1 \wedge u^{(p-1)} + u_1 \wedge d\Gamma^{(p-1)}(A) u^{(p-1)} \,.$$

eigenvalues in $I_A(h) = [0, Ah^{\frac{3}{2}}[)$ is equal to the number of critical points with index 0.

When $p > 0$, the quantity

$$\mu_1 = \sum_{j=1}^{p}(|\lambda_j| + \lambda_j) + \sum_{j=p+1}^{n}(|\lambda_j| - \lambda_j),$$

vanishes only when Hess f admits p negative eigenvalues and $n - p$ positive eigenvalues.

Therefore, the dimension of the eigenspace corresponding to the small eigenvalues of $\Delta_{f,h}^{(p)}$ (i.e. for eigenvalues in $[0, \epsilon_0 h]$, for some sufficiently small $\epsilon_0 > 0$) is equal to the number of points of index p.

We just finish this chapter by sketching what will be analyzed more deeply in the next chapters.

The semi-classical analysis shows that the eigenvalues in $I_A(h)$ are actually exponentially small. The corresponding eigenfunctions of $\Delta_{f,h}^{(0)}$ are localized near the points of index 0, that is near the set of local minima, with exponential decay outside this set. Moreover one can find a basis of the eigenspace corresponding to $I_A(h)$ where each element is localized in one and only one critical point of index 0 (we label the elements of this basis by $\mathcal{C}^{(0)}$).

Indeed the starting point is to observe that $\chi_j \exp -\frac{f}{h}$ where χ_j is a cut-off function localizing near a local minimum U_j of f. The analysis of the Witten Laplacian on the 1-forms shows that the eigenfunctions are localized near the points of index 1 and one can also find a basis of the eigenspace $E_f^{(1)}(I_A(h))$ localized at the points of index 1. We can label the elements of this basis by $\mathcal{C}^{(1)}$.

One way to understand this is to observe, as in Subsection 11.2.2, that if u is an eigenvector for $\Delta_{f,h}^{(p)}$ attached to some eigenvalue λ then $d_{f,h} u$ is either 0 or a non zero eigenfunction of $\Delta_{f,h}^{(p+1)}$ hence (if $\lambda \in I_A(h)$) localized in the union of the points of index 1. By combining this information with those on the decay of the eigenfunctions, one can get that the eigenvalues of $\Delta_{f,h}^{(0)}$ are actually exponentially small.

Semi-classical Analysis and Witten Laplacians: Tunneling Effects

This chapter is devoted to the brief presentation of the analysis of Helffer-Sjöstrand [HelSj4] as initially inspired by Witten [Wi]. The hope was that it would permit to recover the analysis proposed by Bovier-Gayrard-Klein [BovGayKl] by a different approach. This is actually only partially true and this will be discussed in the next chapter.

15.1 Morse Theory, Agmon Distance and Orientation Complex

15.1.1 Morse Function and Agmon Distance

If f is a Morse function, we would like to analyze the relation between the variation of f and the Agmon distance relative to $V = |\nabla f|^2$.
The main points are the following:

Lemma 15.1.
For any x and y on the manifold M, we have:

$$|f(x) - f(y)| \leq d_V(x, y)$$

This is immediate by writing that for a path γ such that $\gamma(0) = x$ and $\gamma(1) = y$, one has

$$f(x) - f(y) = \int_0^1 \nabla f(\gamma(t)) \cdot \gamma'(t)\, dt \ .$$

It is important to understand the cases when one has equality. This is analyzed in the following lemma.

Lemma 15.2.
If $x, y \in M$ and $f(x) - f(y) = d_V(x, y)$ then any minimal geodesic (for the Agmon distance) from y to x is a generalized integral curve of ∇f (we have to be careful at the critical points of f).

Let us describe more explicitly what is a generalized integral curve. This is a continuous curve in M, which, except at a finite number of critical points of f (which could be end points), is an integral curve of ∇f. Between two critical points x_1 and x_2 , one can find a parametrization of the curve

$$] - \infty, +\infty[\ni t \mapsto \gamma(t) ,$$

such that:

$$\gamma'(t) = \nabla f(\gamma(t)) , \ \lim_{t \to -\infty} \gamma(t) = x_1 , \ \lim_{t \to +\infty} \gamma(t) = x_2 .$$

15.1.2 Generic Conditions on Morse Functions

Two generic assumptions (appearing also in the standard Morse theory) are made on the Morse function f for the results obtained by Helffer-Sjöstrand [HelSj4]. The first one is

Assumption 15.3.
If U_j and U_k are two critical points of f and if

$$d(U_j, U_k) = f(U_j) - f(U_k) ,$$

then

$$\ell_j \geq \ell_k + 1 .$$

For example, if U_k is a local minimum we have $\ell_k = 0$, and this will imply $\ell_j \geq 1$.

At each point U_j of index ℓ, one can associate the outgoing stable manifold V_j^+ defined as the union of the trajectories of ∇f starting[1] from U_j and the incoming stable manifold V_j^- defined as the union of the trajectories of ∇f arriving[2] at U_j. V_j^+ is of dimension $n - \ell$ and V_j^- is of dimension ℓ. The stable manifold theorem says that these two sets (outgoing and incoming) are locally manifolds and that their tangent space at U_j are respectively the positive and negative eigenspaces of Hess $f(U_j)$. When $\ell = 0$, V_j^- is empty.

The second generic assumption is

Assumption 15.4.
If $U_k^{(\ell)}$ is a critical point of index ℓ, $U_j^{(\ell+1)}$ is a critical point of index $\ell + 1$ and if

$$d(U_k^{(\ell)}, U_j^{(\ell+1)}) = f(U_j^{(\ell+1)}) - f(U_k^{(\ell)}) ,$$

then there is only a finite number of minimal geodesics from $U_k^{(\ell)}$ to $U_j^{(\ell+1)}$. Moreover V_k^+ and V_j^- intersect transversally along these minimal geodesics.

---

[1] A trajectory "starting from U_j" is a curve γ s.t. $\gamma'(t) = \nabla f(\gamma(t))$ and $\lim_{t \to -\infty} \gamma(t) = U_j$,

[2] A trajectory "arriving at U_j" is a curve γ s.t. $\gamma'(t) = \nabla f(\gamma(t))$ and $\lim_{t \to +\infty} \gamma(t) = U_j$,

Under these assumptions, two cases appear for a given pair of critical points of index ℓ and $\ell + 1$:

$$\text{Case 1:} \qquad d(U_k^{(\ell)}, U_j^{(\ell+1)}) > f(U_j^{(\ell+1)}) - f(U_k^{(\ell)}) , \qquad (15.1)$$

and

$$\text{Case 2:} \qquad d(U_k^{(\ell)}, U_j^{(\ell+1)}) = f(U_j^{(\ell+1)}) - f(U_k^{(\ell)}) . \qquad (15.2)$$

In the second case, the minimal geodesics for the Agmon metric are generalized integral curves of ∇f.

Remark 15.5.
As a consequence of the two previous assumptions, we observe that in Case 2, no minimal geodesic could meet another critical point of f (cf [HelSj4], p. 177).

Remark 15.6.
It is observed in [ZH] (Chapter 5), that these conditions, which are also called "Smale transversality conditions" in reference to [Sm2], are generic.

These assumptions, which will be useful when we will apply the Laplace integral method for computing explicitly the main term, lead in case 2 to the following property.

Property 15.7.
Between two critical points $U_k^{(\ell)}$ and $U_j^{(\ell+1)}$ satisfying (15.2), and for any minimal geodesic γ between these two points the map

$$x \mapsto d_V(U_k^{(\ell)}, x) + d_V(x, U_j^{(\ell+1)}) - d_V(U_k^{(\ell)}, U_j^{(\ell+1)})$$

vanishes exactly at order 2 when restricted, at some point x_γ of γ, to a transversal hypersurface to γ.

15.1.3 Orientation Complex

Under these generic assumptions, one can define the orientation complex in the following way. We assume that M is oriented. We choose at each critical point an orientation of V_k^+ which gives automatically a natural orientation of V_k^-. Now, if γ is a minimal geodesic between $U_k^{(\ell)}$ and $U_j^{(\ell+1)}$, and if we are in Case (2) (cf Subsection 15.1.2), the generic condition says that V_k^+ and V_j^- intersect transversally along generalized integral curves of ∇f. We introduce the index ϵ_γ by comparing at some point x_γ of γ the orientation of $(T_{x_\gamma} V_k^+)^\perp$ and the orientation of the orthogonal of $\nabla f(x_\gamma)$ in $T_{x_\gamma} V_j^-$. (In the case when $(T_{x_\gamma} V_k^+) = \{0\}$, we simply compare the orientation of ∇f and of $T_{x_\gamma} V_j^-$. We then define:

$$\begin{cases} \beta_{jk}^{(\ell)} = \sum_\gamma \epsilon_\gamma & \text{in case 2 ,} \\ \beta_{jk}^{(\ell)} = 0 & \text{in case 1 ,} \end{cases} \qquad (15.3)$$

where the sum $\sum_\gamma$ is over the minimal geodesics between $U_k^{(\ell)}$ and $U_j^{(\ell+1)}$. The orientation complex ∂ is defined as:

$$0 \to \mathbb{C}^{m_0} \xrightarrow{\partial^{(0)}} \mathbb{C}^{m_1} \to \cdots \to \mathbb{C}^{m_\ell} \xmapsto{\partial^{(\ell)}} \mathbb{C}^{m_{(\ell+1)}} \to \cdots \to \mathbb{C}^{m_{(n-1)}} \xrightarrow{\partial^{(n-1)}} \mathbb{C}^{m_n} \to 0 ,$$

(15.4)

where in degree ℓ the matrix is the matrix $\beta_{jk}^{(\ell)}$.
The main theorem in this direction is:

Theorem 15.8.
Under the assumptions (15.3) and (15.4), the Betti numbers are the same as the cohomology numbers of the orientation complex ∂.

The existence of a semi-classical proof of this result was suggested by Witten [Wi] and proved by [HelSj4]. According to [ZH], a topological proof was previously given by F. Laudenbach (appendix in [BZ]). Other results about the connected Bott inequalities are obtained (in the case of degenerate critical points) in [HelSj5] (see also [Bi1] and [Hel5]).

15.2 Semi-classical Analysis of the Witten Laplacians

We follow here rather closely [HelSj4] but with additional remarks about the particular case $\ell = 0$. Let us come back to the various steps of the analysis.

15.2.1 One Well Reference Problems

We denote by $M_j^{(\ell)}$ the open set obtained by substracting balls of sufficiently small radius η around the other points of index ℓ, $U_k^{(\ell)}$ $(k \neq j)$:

$$M_j^{(\ell)} := M \setminus \cup_{k \neq j} B(U_k^{(\ell)}, \eta) .$$

Attached to each of these Dirichlet problems there is a ground state energy $\mu_j^{(\ell)}(h)$ and a corresponding ground state $\phi_j^{(\ell)}$ with the following decay in the $H^1(M_j^{(\ell)})$-norm:

$$\phi_j^{(\ell)} = \tilde{O}(\exp -\frac{d(x, U_j^{(\ell)})}{h}) .$$

(15.5)

Here $\tilde{O}(f)$ (for a non negative f) means $\mathcal{O}_\epsilon(f \exp \frac{\epsilon}{h})$ for any $\epsilon > 0$. This means that we loose multiplicatively $\mathcal{O}_\epsilon(\exp \frac{\epsilon}{h})$ but for an arbitrarily small $\epsilon > 0$. This result is based on Agmon estimates. It is already not a trivial result, in the sense that one uses that the wells of index different of ℓ are non resonant (see the discussion around (13.32) including the footnote). The proof given in Section 13.4 would only have given the weaker estimate:

$$\phi_j^{(\ell)} = \tilde{O}(\exp -\frac{\hat{d}_j(x)}{h}) \,, \tag{15.6}$$

with

$$\hat{d}_j(x) = d(x, |\nabla f|^{-1}(0) \cap M_j^{(\ell)}) \,.$$

In order to have a function defined globally, we introduce a cut-off function χ_j with supp χ_j in $M_j^{(\ell)}$ but equal to one in $M \setminus \cup_{k \neq j} B(U_k^{(\ell)}, 2\eta)$. If we introduce

$$\psi_j^{(\ell)} = \chi_j \phi_j^{(\ell)} \,,$$

these functions span a space which gives a good approximation of the eigenspace corresponding to the small eigenvalues of $\Delta_{f,h}^{(\ell)}$ with some error of order $\tilde{O}\left(\exp -\frac{1}{h} \min_{j \neq k} d(U_j^{(\ell)}, U_k^{(\ell)})\right)$.

15.2.2 Improved Decay

Here are improvements established in [HelSj4] for the decay of $d_{f,h}\phi_j^{(\ell)}$ and $d_{f,h}^*\phi_j^{(\ell)}$. They are obtained by combining the information on the decay of $\phi_j^{(\ell)}$ and the information on the decay of $d_{f,h}\phi_j^{(\ell)}$ as an eigenfunction (if not 0) of the Witten Laplacian on the $(\ell+1)$-forms and of $d_{f,h}^*\phi_j^{(\ell)}$ as an eigenfunction (if not 0) of the Witten Laplacian on the $(\ell-1)$-forms. We have indeed

$$d_{f,h}\phi_j^{(\ell)} = \tilde{O}(\exp -\frac{\alpha_j^\ell(x)}{h}) \,, \tag{15.7}$$

where

$$\alpha_j^\ell(x) = \min_{k \in \mathcal{C}^{(\ell+1)} \cup \mathcal{C}^{(\ell)} \setminus \{j\}} d(U_j^{(\ell)}, U_k) + d(U_k, x) \,. \tag{15.8}$$

Similarly, we get:

$$d_{f,h}^*\phi_j^{(\ell)} = \tilde{O}(\exp -\frac{\beta_j^\ell(x)}{h}) \,, \tag{15.9}$$

where

$$\beta_j^\ell(x) = \min_{k \in \mathcal{C}^{(\ell-1)} \cup \mathcal{C}^{(\ell)} \setminus \{j\}} d(U_j^{(\ell)}, U_k) + d(U_k, x) \,. \tag{15.10}$$

As a corollary, we get

Lemma 15.9.
The first eigenvalue $\mu_j^{(\ell)}(h)$ satisfies

$$\mu_j^{(\ell)} = \tilde{O}(\exp -\frac{2c_j^{(\ell)}}{h}) \,, \tag{15.11}$$

with

$$c_j^{(\ell)} = \min_{k \in \mathcal{C}^{(\ell-1)} \cup \mathcal{C}^{(\ell)} \cup \mathcal{C}^{(\ell+1)} \setminus \{j\}} d(U_j^{(\ell)}, U_k) \,. \tag{15.12}$$

When $\ell = 0$, we get that $c_j^{(\ell)}$ is the minimal distance between $U_j^{(0)}$ and a saddle point of index 1.

This can be reobtained more easily by considering directly a cut-off function $\tilde{\chi}_j$ and considering $\tilde{\chi}_j \exp -\frac{f}{h}$.

15.2.3 An Adapted Basis

Once we have constructed the $\psi_j^{(\ell)}$, one can then take their projection onto the eigenspace $F := E_f^{(\ell)}(I_A(h))$, attached to the eigenvalues in $I_A(h) = [0, Ah^{\frac{3}{2}}[$ and consider:

$$v_j^{(\ell)} = \Pi_F \psi_j^{(\ell)} .$$

One can prove (see Lemma 13.6) that $v_j^{(\ell)} - \psi_j^{(\ell)}$ is exponentially small in L^2. Let us consider the matrix

$$V_{jk} := \langle v_j^{(\ell)} \mid v_k^{(\ell)} \rangle .$$

This matrix is exponentially close to the identity. Orthonormalizing by the Gram-Schmidt procedure, that is considering

$$e_j^{(\ell)} = \sum_{k=1}^{m_\ell} (V^{-\frac{1}{2}})_{jk} v_k^{(\ell)} , \quad \text{for } j = 1, \dots, m_\ell ,$$

leads to the orthonormal basis $(e_j^{(\ell)})$ of this eigenspace .
The analysis gives that, for the computations of the main term, it is enough to work with the $v_j^{(\ell)}$ modulo some controlled error and that

$$v_j^{(\ell)} - \psi_j^{(\ell)} = \tilde{O}(\exp -\frac{\delta_j^{(\ell)}(x)}{h}), \tag{15.13}$$

with

$$\delta_j^{(\ell)}(x) = \min_{k \in \mathcal{C}^{(\ell)} \setminus \{j\}} \left(d(U_j^{(\ell)}, U_k^{(\ell)}) + d(U_k^{(\ell)}, x) \right) . \tag{15.14}$$

Here the reader should observe that the proof of Lemma 13.6 provides a smaller $\delta_j(x)$ where the minimum would have been considered over all critical points except $U_j^{(\ell)}$.

15.2.4 WKB Approximation

We already discussed the existence of WKB solutions in Section 13.6. The fact that we are working with systems does not lead to very difficult new problems. In suitable[3] domains one can consequently approximate $v_j^{(\ell)}$ by a

---

[3] geodesically starshape with respect to U_j

WKB solution. When $\ell = 0$, the function $\exp -\frac{f(x)}{h}$ is after renormalization around the local minima $U_j^{(0)}$ the WKB approximation. For the point of index 1, we shall see that we only need the WKB approximation in a neighborhood of the saddle points. This construction is done in [HelSj4].

As observed in [HelSj4], all this material does not give the way to analyze directly the Witten Laplacian. It is better to first analyze the Witten complex and then to come back to the Witten Laplacian.

15.3 Semi-classical Analysis of the Witten Complex

In the previous section, we have seen that there exists a basis of the space $E_f^{(0)}(I_A(h))$ indexed by $\mathcal{C}^{(0)}$ and a basis of the space of $E_f^{(1)}(I_A(h))$ indexed by $\mathcal{C}^{(1)}$, consisting of exponentially localized elements and admitting good WKB approximations. Moreover we have seen that $d_{f,h}^{(0)}$ sends $E_f^{(0)}(I_A(h))$ into $E_f^{(1)}(I_A(h))$ and is represented in the above mentioned basis by a $m_1 \times m_0$ matrix.

This can actually be done at each degree and the computation of this matrix (up to a well controlled remainder estimated by Helffer-Sjöstrand [HelSj4]) permits to recover as a limiting complex the so called orientation complex (see for example [ZH] and Subsection 15.1.3). Here we will limit ourself to the analysis on 0-forms.

We denote by

$$M = (M_{jk})_{\{j=1,\dots,m_1 \,,\, k=1,\dots m_0\}}$$

the matrix of $d_{f,h}^{(0)} {}_{/E_f^{(0)}(I_A(h))}$ in the above basis of $E_f^{(0)}(I_A(h))$ and $E_f^{(1)}(I_A(h))$. The main term to compute is

$$M_{jk} = \langle e_j^{(1)} \mid d_{f,h} e_k^{(0)} \rangle . \tag{15.15}$$

We observe that:

$$M_{jk} = I_{jk} + \mathcal{O}\left(\exp -\frac{(d(U_j^{(1)}, U_k^{(0)}) + \alpha_{jk})}{h} \right) , \tag{15.16}$$

for some $\alpha_{jk} > 0$, with

$$I_{jk} = \langle v_j^{(1)} \mid d_{f,h} v_k^{(0)} \rangle , \tag{15.17}$$

which is easier to compute. We follow [HelSj4] but change a little the cut-off function. This choice can be useful for other purpose.

We introduce a cut-off function $\hat{\chi}_j$ with the property that $\chi_j = 1$ in a neighborhood of $U_j^{(1)}$ say a ball of size $\eta_j > 0$ and with compact support in $B(U_j^{(1)}, 2\eta_j)$. This η_j should be larger than the "cut-off" appearing previously in the construction of $\psi_k^{(0)}$. In [HelSj4], the cut-off was more or less at the middle between $U_j^{(1)}$ and $U_k^{(0)}$. Here we prefer to use that we have a better information on $\psi_k^{(0)}$.

We now show that this other choice does not modify the estimate of [HelSj4]. We first decompose I_{jk} as the sum

$$I_{jk} = I_{jk}^1 + I_{jk}^2 ,$$

with

$$I_{jk}^1 = \langle (1 - \hat{\chi}_j) v_j^{(1)} \mid d_{f,h} v_k^{(0)} \rangle .$$

Using (15.5) and (15.7), we obtain that the term $((1-\hat{\chi}_j) v_j^{(1)} \cdot d_{f,h} v_k^{(0)})(x)$ (scalar product of two one-forms at x) which is integrated for giving I_{jk}^1 behaves like

$$\tilde{\mathcal{O}} \left(\exp -\frac{1}{h} \left(d(x, U_j^{(1)}) + \min_{p \in \mathcal{C}^{(0)} \cup \mathcal{C}^{(1)} \setminus \{k\}} (d(U_p, x) + d(U_p, U_k^{(0)})) \right) \right) .$$

Hence, by taking into account the information on the support of $(1 - \hat{\chi}_j)$, it is of order $\mathcal{O}(\exp -\frac{1}{h}(d(U_j^{(1)}, U_k^{(0)}) + \alpha_j))$ for $\alpha_j > 0$.

Let us now treat the second term. We first rewrite:

$$I_{jk}^2 = \langle d_{f,h}^* v_j^{(1)} \mid \hat{\chi}_j v_k^{(0)} \rangle - h \langle d\hat{\chi}_j \wedge v_k^{(0)} \mid v_j^{(1)} \rangle .$$

Using the estimates (15.5) and (15.9), the first term of I_{jk}^2 can be treated similarly. So we finally get the following lemma.

Lemma 15.10.
There exists a constant $\alpha_j > 0$ such that

$$I_{jk} = -h \langle d\hat{\chi}_j \wedge v_k^{(0)} \mid v_j^{(1)} \rangle + \mathcal{O} \left(\exp -\frac{1}{h}(d(U_j^{(1)}, U_k^{(0)}) + \alpha_j) \right) \qquad (15.18)$$

We are reduced to get a good knowledge of $v_k^{(0)}$ and $v_j^{(1)}$ on the support of $d\hat{\chi}_j$ which is contained in $B(U_j^{(1)}, 2\eta_j) \setminus B(U_j^{(1)}, \eta_j)$. A further analysis involving only Agmon estimates shows that we should also restrict the approximation to a neighborhood $\mathcal{V}_{kj}$ of the minimal geodesic between $U_j^{(1)}$ and $U_k^{(0)}$.

The new point (in the case $\ell = 0$) is that there $v_k^{(0)}$ is well approximated by $\rho_k(h) \exp -\frac{1}{h}(f(x) - f(U_k^{(0)}))$, in $\mathcal{V}_{kj} \cap \{B(U_j^{(1)}, 2\eta_j) \setminus B(U_j^{(1)}, \eta_j)\}$. The proof of this point can be done following the proof of the so-called "analytic case" of Theorem 5.8 in [HelSj1].

According to our choice of $\hat{\chi}_j$, we also need to know the WKB approximation of $v_j^{(1)}$ in the ball $B(U_j^{(1)}, 2\eta_j)$. This construction, corresponding to an extension to systems of what we sketched in Section 13.6 and Subsection 15.2.4 is given in [HelSj4]. Again the comparison between the WKB construction and $v_j^{(1)}$ is obtained by using the "C^∞" part of Theorem 5.8 in [HelSj1].

16

Accurate Asymptotics for the Exponentially Small Eigenvalues of $\Delta_{f,h}^{(0)}$

The aim of this chapter is to present the results obtained by Bovier-Gayrard-Klein [BovGayKl] and recent generalizations which we have obtained in collaboration with M. Klein [HelKlNi]. After the description of the main results, we will try to analyze how far we can go with the Helffer-Sjöstrand techniques. This will lead us to give a relatively selfcontained proof in the case when f has two local minima and one unique saddle point.

16.1 Assumptions and Labelling of Local Minima

The labelling of the local minima is essentially the one introduced by Bovier-Gayrard-Klein [BovGayKl]. It is an important point of their probabilistic approach and their intuition was based on the notion of exit times for the stochastic dynamics. Their idea was to enumerate the local minima according to the decreasing order of exit times. These authors work in the case of $\Omega = \mathbb{R}^n$. The cases when M is a compact manifold or a bounded regular open set are analyzed in [HelKlNi] and [HelNi2].

One key notion is the notion of non degenerate saddle point between two sets A and B in $\mathbb{R}^n$. First we define

$$H_f(A, B) = \inf_{\omega; \omega(0) \in A,\ \omega(1) \in B}\ \sup_{t \in [0,1]} f(\omega(t)) , \qquad (16.1)$$

where the infimum is over all continuous paths going from A to B. We then say that $z^* = z^*(A, B)$ is a non degenerate saddle point between A and B if:

•

$$f(z^*(A, B)) = H_f(A, B) ; \qquad (16.2)$$

• $z^*(A, B)$ is a critical point of index 1 ;

- There exists a continuous path $[0,1] \ni t \mapsto \omega(t)$ going from A to B and $t_0 \in]0,1[$ such that:

$$z^*(A,B) = \omega(t_0) \text{ and } f(\omega(t)) \leq f(z^*(A,B)) \, , \, \forall t \in [0,1] \, .$$

Note that $z^*(A,B)$ is not uniquely defined and we call $Z(A,B)$ the set of all non degenerate saddle points between A and B. The next assumptions ensure that for two disjoint sets A and B of local minima, $Z(A,B)$ is not empty and finite. In order to control the situation at ∞, We assume here that $f \in \mathcal{C}^\infty$ and that there exists a compact K and a constant $C_0 > 0$ such that:

$$|\nabla f| \geq \frac{1}{C_0} \, , \tag{16.3}$$

outside K and that:

$$|\Delta f| \leq C_0 < \nabla f >^2 \, . \tag{16.4}$$

Then we assume that f is locally a Morse function. So there is only a finite number of non degenerate critical points. We note that, under the assumptions (16.3) and (16.4), and using Persson's Theorem, the essential spectrum of $\Delta_{f,h}^{(0)}$ is bounded from below by $\frac{1}{2C_0}$, for h small enough.

Up to now the assumptions leave open the possibility of having $\exp -\frac{f}{h}$ in the kernel or not. In [BovGayKl], the assumption is that

$$\int_{\{y:\, f(y) \geq a\}} \exp(-\frac{f(y)}{h}) dy \leq C \exp -\frac{a}{h} \, ,$$

for any $a \in \mathbb{R}$. This assumption is removed in [HelKlNi] and [HelNi2]. Here, we assume for simplicity:

$$f(x) \geq \frac{1}{C} \langle x \rangle^{\frac{1}{C}} \, . \tag{16.5}$$

which in particularly implies that:

$$\exp -\frac{f}{h} \in L^2(\mathbb{R}^n) \, .$$

We first start with m_0 unlabelled local minima $U_\alpha^{(0)}$, $\alpha \in A$, $\#A = m_0$. In order to give a more constructive presentation of the proper numbering of local minima which is associated with the crucial assumption, we introduce the following initial assumption which is slightly stronger than what is assumed in [BovGayKl][HelKlNi][HelNi2].

Assumption 16.1.
The function f is a Morse $\mathcal{C}^\infty(\mathbb{R}^n)$ function which satisfies (16.4) (16.5) and the following property:
For any $A' \subset A$, $A' \neq A$ and any $\alpha \in A \setminus A'$, there is a unique non-degenerate saddle point $z^*(U_\alpha^{(0)}, \cup_{\alpha' \in A'} U_{\alpha'}^{(0)})$.

The local minima are then indexed according to following induction process on k starting from $k = m_0$:
By assuming that $U_{m_0}^{(0)}, \ldots, U_{k+1}^{(0)}$ are known, one sets

$$A_k = \left\{ \alpha \in A, U_\alpha^{(0)} \notin \left\{ U_{m_0}^{(0)}, \ldots, U_{k+1}^{(0)} \right\} \right\}.$$

The local minimum $U_k^{(0)}$ is then chosen as a minimizer of

$$f(z^*(U_\alpha^{(0)}, \bigcup_{\alpha' \in A_k \setminus \{\alpha\}} U_{\alpha'}^{(0)})) - f(U_\alpha^{(0)}), \quad \alpha \in A_k. \tag{16.6}$$

Remark 16.2.
In our case, once $U_2^{(0)}$ is known, only one local minimum $U_1^{(0)}$ is left and it has to be a global minimum for f.

Here is the crucial assumption of [BovGayKl] which avoids to have exponentially small eigenvalues with close asymptotics[1] .

Assumption 16.3.
At each step of the previous induction process, the quantity (16.6) has a unique minimizer ($U_k^{(0)}$ is uniquely defined).

Once the local minima are labelled, one can fix the critical points of index 1 which play an important role:

$$\forall k \in \{2, \ldots, m_0\}, \quad U_{j(k)}^{(1)} = z^*(U_k^{(0)}, \cup_{j \in \{1, \ldots, k-1\}} U_j^{(0)}). \tag{16.7}$$

For $k = 1$ we set $j(1) = 0$ where $U_0^{(1)}$ does not denote a point of $\mathbb{R}^n$ (it can be interpreted as $U_0^{(1)} = \infty$) with the convention $f(U_0^{(1)}) = +\infty$.
Note that the assumption 16.3 implies that the quantities

$$f(U_{j(k)}^{(1)}) - f(U_k^{(0)}), \quad k \in \{1, \ldots, m_0\} \tag{16.8}$$

are strictly decreasing.

16.2 Main Result

The main result, we would like to discuss is:

Theorem 16.4.
If f satisfies the Assumptions 16.1 and 16.3 and if the local minima $U_1^{(0)}, \cdots, U_{m_0}^{(0)}$ are labelled according to the previous process with $U_{j(k)}^{(1)}$ defined by (16.7) then

---

[1] It will avoid the existence of two eigenvalues $\lambda(h)$ and $\lambda'(h)$, with corresponding orthogonal eigenfunctions, such that $\lim_{h \to 0} h \log \lambda(h) = \lim_{h \to 0} h \log \lambda'(h)$

there exists $h_0 > 0$, such that, for $h \in]0, h_0]$, the m_0 (exponentially small) lowest eigenvalues satisfy

$$\lambda_k(h) = \frac{h}{\pi} |\widehat{\lambda}_1(U_{j(k)}^{(1)})| \sqrt{\left| \frac{\det(\text{Hess } f(U_k^{(0)}))}{\det(\text{Hess } f(U_{j(k)}^{(1)}))} \right|}$$
$$\times \exp{-\frac{2}{h} \left(f(U_{j(k)}^{(1)}) - f(U_k^{(0)}) \right)} \times (1 + r_k(h)) ,$$

(16.9)

where $\widehat{\lambda}_1(U_j^{(1)})$ denotes the unique negative eigenvalue of the Hessian of f at the saddle point $U_j^{(1)}$ and $r_k(h) = o(1)$.

Remarks 16.5.

1. One has to give the interpretation of the formula for the first eigenvalue. Then one knows $\lambda_1(h) = 0$ and it is coherent with our convention

$$f(U_{j(1)}^{(1)}) = +\infty ,$$

(16.10)

   with any finite value for the other factors. One observes that $U_1^{(0)}$ corresponds to a global minimum.
2. This theorem is proved with

$$r_k(h) = \mathcal{O}(h^{\frac{1}{2}} |\ln h|) ,$$

   by Bovier-Gayrard-Klein [BovGayKl], with slightly weaker assumptions and with

$$r_k(h) \sim \sum_{j \geq 1} r_{jk} h^j ,$$

   by Helffer-Klein-Nier [HelKlNi].
3. The probabilistic approach of [BovGayKl] permits a lower regularity for f ($f \in \mathcal{C}^3$).
4. The approach in [HelKlNi] works for situations where the probabilistic approach fails or presents additional difficulties:
   - some cases when $\Omega = \mathbb{R}^n$ and $e^{-f(x)/h} \notin L^2$,
   - Ω is a compact oriented Riemannian manifold.
5. The question of the splitting between 0 and the lowest non zero eigenvalue has been attacked before by other authors including for example Holley-Kusuoka-Strook [HolKusStr], Miclo [Mi] and Bovier-Eckhoff-Gayrard-Klein [BovEckGayKl1], [BovEckGayKl2].

16.3 Proof of Theorem 16.4 in the Case of Two Local Minima

We note that the corresponding matrix of $\Delta_{f,h}^{(0)}{}_{/E_f^{(0)}(I_A(h))}$ is $\mathcal{M}^* \circ \mathcal{M}$. This matrix has 0 as an eigenvalue: the function $\exp{-\frac{f}{h}}$ is indeed in the kernel of

$\Delta_{f,h}^{(0)}$. One is reduced to the analysis of $\mathcal{M}^* \circ \mathcal{M}$ and of its approximation. At a first glance, a good knowledge of this matrix, as it results from [HelSj4], seems to contain all the information. It is actually not the case !! In the case when there is one global minimum and one other local minimum, which leads to a 2×2 matrix, with 0 as eigenvalue, it is immediate to recover the second one by taking the asymptotic expansion of the trace of this 2×2 matrix. Nevertherless, when there are more than two local minima, one is only able to compute the largest of these exponentially small eigenvalues lying in $[0, Ah^{\frac{3}{2}}[$, without adding additional information.

We now consider the case with to local minima. Hence we add the assumption that f has two unequal minima $U_1^{(0)}$ and $U_2^{(0)}$ ($U_1^{(0)}$ being the global minimum) and one saddle point $U^{(1)}$ of index 1. We assume that the pairs $(U_1^{(0)}, U^{(1)})$ and $(U_2^{(0)}, U^{(1)})$ correspond to the case (2). In this case, we have:

$$d(U_1^{(0)}, U^{(1)}) = f(U^{(1)}) - f(U_1^{(0)}),$$
$$d(U_2^{(0)}, U^{(1)}) = f(U^{(1)}) - f(U_2^{(0)}),$$
$$d(U_1^{(0)}, U_2^{(0)}) = d(U_1^{(0)}, U^{(1)}) + d(U^{(1)}, U_2^{(0)}) = 2f(U^{(1)}) - f(U_1^{(0)}) - f(U_2^{(0)}).$$

The matrix $\mathcal{M}$ is simply:

$$\mathcal{M} = (\beta_1, \beta_2)$$

where the coefficients can be estimated by comparison with WKB constructions:

$$\beta_j = h^{\nu_j} a_j(h) \exp{-(f(U^{(1)}) - f(U_j^{(0)}))/h},$$

where $a_j(h)$ has a complete expansion in powers of h. There is a more accurate formula, which was important for the proof of Theorem 15.8 and which is given in [HelSj4] (see (3.24) and (3.27) therein), that we recall now. If j is a point of index ℓ and k is a point of index $\ell + 1$, the formula is (in case 2)

$$M_{jk} = (\frac{h}{\pi})^{\frac{1}{2}} (\sum \epsilon_\gamma + \mathcal{O}(h)) \Lambda_j \Lambda_k^{-1} \exp{-(f(U_k^{(\ell+1)}) - f(U_j^{(\ell)}))/h}, \quad (16.11)$$

where at a critical point $U_p^{(m)}$ of index m

$$\Lambda_p = \left(\frac{|\widehat{\lambda}_{m+1}| \cdots |\widehat{\lambda}_n|}{|\widehat{\lambda}_1| \cdots |\widehat{\lambda}_m|} \right)^{\frac{1}{4}}, \quad (16.12)$$

and the $\widehat{\lambda}_q$'s (more precisely $\widehat{\lambda}_q(U_p)$) are the eigenvalues arranged in increasing order, such that, in particular, $\widehat{\lambda}_q < 0$ for $q \leq m$ and $\widehat{\lambda}_q > 0$ for $q > m$, of Hess $f(U_p^{(m)})$.

In case 1, one can only show that

$$M_{jk} = \mathcal{O}(\exp{-d(U_j^{(\ell)}, U_k^{(\ell+1)})/h}). \quad (16.13)$$

In our particular case, we get, observing the uniqueness of the minimal geodesic between the saddle point and the local minimum:

$$\beta_j =$$
$$\frac{1}{\sqrt{\pi}} h^{\frac{1}{2}} \left(\epsilon_j + \mathcal{O}(h)\right) \left(|\widehat{\lambda}_1(U_j^{(0)})| \cdots |\widehat{\lambda}_n(U_j^{(0)})|\right)^{\frac{1}{4}} \left(\frac{|\widehat{\lambda}_2(U^{(1)})|\cdots|\widehat{\lambda}_n(U^{(1)})|}{|\widehat{\lambda}_1(U^{(1)})|}\right)^{-\frac{1}{4}}$$
$$\exp-(f(U^{(1)}) - f(U_j^{(0)}))/h \,,$$

$$(16.14)$$

with $|\epsilon_j| = 1$.

Remark 16.6.
It could appear to be strange to see ϵ_j appearing in (16.14), but the choice of the basis attached to the critical points of index 1 is not canonical, and related to the choice of an orientation of the "positive spaces". In any case, this is the square of β_j which will be relevant for the computation of the eigenvalues.

Remark 16.7.
Note that this result is improved in [HelSj4] (Theorem 6.1) by a better control of $\mathcal{O}(h)$ in (16.14).

The matrix of the "reduced" Witten Laplacian is then:

$$\mathcal{M}^*\mathcal{M} = \begin{pmatrix} \beta_1^2 & \beta_1\beta_2 \\ \beta_1\beta_2 & \beta_2^2 \end{pmatrix} \,.$$

We observe that the trace is given by:

$$\mathrm{Tr}\,\mathcal{M}^*\mathcal{M} = \beta_1^2 + \beta_2^2 \,,$$

and that, as expected, 0 is an eigenvalue (compute the determinant!). The non zero eigenvalue λ_2 is given by:

$$\lambda_2 = \beta_1^2 + \beta_2^2 \,.$$

We get consequently in this particular case a complete expansion of λ_2, whose principal term is given by the principal term of β_2^2.
In this particular case, this confirms and completes the results of [BovGayKl].

The case of many saddle points (satisfying case 2) (say p saddle points of index 1 and still two local minima) is treated in the same way. The computation of the trace gives now:

$$\mathrm{Tr}\,\mathcal{M}^*\mathcal{M} = \sum_{k=1}^{p}(M_{k1}^2 + M_{k2}^2) \,.$$

We note that 0 is no more an evident eigenvalue of $\mathcal{M}^*\mathcal{M}$. On the other hand we know that $\exp-\frac{f}{h}$ is in the kernel of d_f and this induces relations. Written in the adapted basis, we have a natural element (α_1, α_2) in the kernel of $\mathcal{M}$. This implies the relations:

$$M_{k1}M_{k'2} = M_{k'1}M_{k2} \ , \ \forall k, k' \ . \tag{16.15}$$

Let us show how to get an asymptotics of α_j. The L^2-renormalization of $\exp -\frac{f}{h}$ leads to $\alpha \exp -\frac{f}{h}$ with α given by the Laplace integral method. We get then $\alpha_1 \sim 1$. For α_2, we have to compute the scalar product of $\alpha \exp -\frac{f}{h}$ with the renormalized $\alpha^2 \chi_2(x) \exp -\frac{f-f(U_2^{(0)})}{h}$.

Another way to understand the structure is to see from (16.14) that M_{jk} has the structure $\delta_j \delta_k^{-1}$ at the level of the principal symbol.

The main contribution will be given by the lowest saddle point $U_{k_{min}}^{(1)}$ corresponding to the smallest $d(U_2^{(0)}, U_k^{(1)}) = f(U_k^{(1)}) - f(U_2^{(0)})$. So the main order is $\exp -2 \inf_k (f(U_{k_{min}}^{(1)}) - f(U_2^{(0)}))/h$.

More precisely, we have obtained for the second eigenvalue:

$$\begin{aligned}
\lambda_2(h) \\
= \frac{h}{\pi}(1 + \mathcal{O}(h))(|\det \mathrm{Hess}\, f(U_2^{(0)})|)^{\frac{1}{2}}(|\det \mathrm{Hess}\, f(U_{k_{min}}^{(1)})|)^{-\frac{1}{2}}|\hat{\lambda}_1(U_{k_{min}}^{(1)})| \\
\times \exp -\frac{2}{h}(f(U_{k_{min}}^{(1)}) - f(U_2^{(0)})) \ .
\end{aligned}$$
$$\tag{16.16}$$

This proves in this particular case Theorem 16.4 with a complete expansion of the prefactor. The method can be extended without particular problems for the case of $m_0 > 2$ local minima but will only lead to a an expansion for the corresponding λ_{m_0}. In particular, we do not get the most important information about the splitting when $m_0 > 2$.

16.4 Towards the General Case

Let us describe some of the ideas involved in [HelKlNi]. As mentioned above in the case $m_0 > 2$, one can always an accurate approximation for the highest eigenvalue $\lambda_{m_0}(h)$ of $\mathcal{M}^*\mathcal{M}$, since $\lambda_{m_0}(h) = \|\mathcal{M}\|^2$. By using the spectral theorem, one can get an accurate information about the corresponding eigenvector as well. Nevertheless a standard orthonormalization process cannot be used here because it brings error terms which are bigger than the other eigenvalues which we are looking for.

The solution which is proposed in [HelKlNi] relies on the fact that the eigenvalues of $\mathcal{M}^*\mathcal{M}$ are the square of the singular values of $\mathcal{M}$:

$$\forall k \in \{1, \ldots, m_0\}, \ \lambda_k(h) = \mu_{m_0+1-k}(\mathcal{M})^2 \ .$$

Without specifying any basis, it reads

$$\forall k \in \{1, \ldots, m_0\}, \ \lambda_k(h) = \mu_{m_0+1-k}(\beta_{f,h}^{(0)})^2 \ ,$$

where $\beta_{f,h}(0)$ is the restricted differential $\beta_{f,h}^{(0)} = d_{f,h}^{(0)}|_{F^{(0)}}$:

$$\beta_{f,h}^{(0)} : \; 1_{[0,h^{3/2})}(\Delta_{f,h}^{(0)}) = F^{(0)} \xrightarrow{d_{f,h}^{(0)}} F^{(1)} = 1_{[0,h^{3/2}]}(\Delta_{f,h}^{(1)}) \; .$$

The important property of the singular eigenvalues $\mu_j(\beta_{f,h}^{(0)})$, which permits to perform the induction comes from the Fan inequalities (see for example [Sim3]):

$$\mu_j(AB) \le \mu_j(A) \, \|B\| \quad ; \quad \mu_j(BA) \le \|B\| \, \mu_j(A) \; .$$

Actually, if B_ℓ , $\ell = 0, 1$, are linear applications $B_\ell : F^{(\ell)} \to F^{(\ell)}$ such that $\max\left\{ \|B_\ell\|, \|B_\ell^{-1}\| \right\} \le 1 + \rho$ then

$$\forall j \in \{1, \ldots, m_0\}, \quad \frac{\mu_j(B_1 \beta_{f,h}^{(0)} B_0)}{(1+\rho)^2} \le \mu_j(\beta_{f,h}^{(0)}) \le \mu_j(B_1 \beta_{f,h}^{(0)} B_0)(1+\rho)^2 \; .$$

This means that small changes of bases in $F^{(0)}$ and $F^{(1)}$ lead to a small relative error of ALL singular values.

The analysis starts by choosing quasimodes for $\Delta_{f,h}^{(0)}$ and $\Delta_{f,h}^{(1)}$, respectively associated with every local minimum and every critical point with index 1. According to the discussion in Section 15.3, the good quasimodes for $\Delta_{f,h}^{(0)}$ simply take the form $C_h \chi(x) e^{-\frac{f(x)}{h}}$, where the cut-off function χ can be modelled on the level sets of f. The quasimodes for $\Delta_{f,h}^{(1)}$ are given by a local WKB approximation.
We refer the reader to [HelKlNi] for details.

Application to the Fokker-Planck Equation

In [HerNi], it was proved that the rate of exponential return to equilibrium for the equation

$$
\begin{cases}
\partial_t F + v \cdot \partial_x F - \frac{1}{m}\partial_x V(x) \cdot \partial_v F - \frac{\gamma_0}{m\beta}\left(\partial_v - \frac{m\beta}{2}v\right) \cdot \left(\partial_v + \frac{m\beta}{2}v\right) F = 0 , \\
F(x, v, t = 0) = F_0(x, v) ,
\end{cases}
$$

in $L^2(\mathbb{R}^{2n})$ can be estimated quantitatively in terms of m, γ_0, β. In the case of low temperature $\beta = \frac{1}{T} \to +\infty$, their result which holds for general non necessarily Morse functions was not very accurate in the presence of more than one local minimum in the potential $V(x)$. This can be improved with Theorem 16.4.

By assuming $V \in S(\langle x \rangle^{2\mu}, \frac{dx^2}{\langle x \rangle^2})$ and $\langle \nabla V(x) \rangle$ elliptic in the symbol class $S(\langle x \rangle^{2\mu-1}, \frac{dx^2}{\langle x \rangle^2})$, $\mu > 1/2$, (see [HerNi] and Chapter 5 for more general assumptions), it is proved in [HerNi] that the exponential rate τ of exponential return to to the equilibrium[1] satisfies

$$
\frac{\min\{1, \omega_1(V_\beta)\}}{64(5 + 3\gamma_0\sqrt{m}\beta^{\frac{\mu-1}{2\mu}} + 3C_{V_\beta})^2} \leq \tau \leq c_V \sqrt{\omega_1(V_\beta)} \log\left[Q(\sqrt{m}\gamma_0, \beta, \omega_1(V_\beta)\right] .
$$

$$(17.1)$$

The quantities involved in these estimates are:

i) the potential V_β given by $V_\beta(x) = \beta V(\beta^{-1/2\mu}x)$,

ii) the first non zero eigenvalue $\omega_1(V_\beta)$ of the Witten Laplacian $\Delta^{(0)}_{V_\beta/2}$,

iii) the non negative constant C_{V_β} defined by

$$
C_{V_\beta}^2 := \max\left\{\sup_{x\in\mathbb{R}^n} \sigma\left[(\operatorname{Hess} V_\beta)^2 - \left(\frac{1}{4}|\partial_x V_\beta|^2 - \frac{1}{2}\Delta V_\beta\right)\operatorname{Id}\right], 0\right\} .
$$

$$(17.2)$$

---

[1] Note that in the derivation of quantitative estimates in [HerNi] the exponential in time law, $e^{-\tau t}$ is corrected by some rational prefactor $(t + t^{-1})^N$.

iv) the constant c_V which is independent of the physical parameters m, γ_0 and β,

v) the rational function

$$(\gamma, \beta, \omega) \mapsto Q(\gamma, \beta, \omega) := (\gamma + \gamma^{-1})^{N_0} (\beta + \beta^{-1})^{N_0} (\omega + \omega^{-1})^{N_0} ,$$

where $N_0 \in \mathbb{N}$ is universal.

Note that, under the ellipticity assumption, one has the upper bound

$$C_{V_\beta} \le c_V'(1 + \beta^{\frac{\mu-1}{\mu}}),$$

where c_V' is independent of the physical parameters m, γ_0 and β.

In order to apply Theorem 16.4, we also assume that:
V is a Morse function such that $V(x) \le C^{-1}\langle x \rangle^{2\mu}$ and $\nabla V(x) \ne 0$ for $|x| \ge C$.
The quantity $\omega_1(V_\beta)$ is the splitting associated with the Witten Laplacian

$$\Delta_{V_\beta/2}^{(0)} = -\Delta_x + \frac{\beta^{2-2/\mu}}{4}\left|\nabla V(\beta^{-1/2\mu}x)\right|^2 - \frac{\beta^{-2/\mu}}{2}\Delta V(\beta^{-1/\mu}x).$$

After a unitary transform the spectral analysis for $\beta \to \infty$ is the one of the semiclassical operator

$$h^{-2+1/\mu}\left(-h^2\Delta + \frac{1}{4}|\nabla V(x)|^2 - \frac{1}{2}\Delta V(x)\right) = h^{-2+1/\mu}\Delta_{V/2,h}^{(0)} ,$$

where $h = \frac{1}{\beta}$ is the semi-classical parameter.

We consider the case with two minima $U_1^{(0)}$ and $U_2^{(0)}$ and assume that $U_1^{(0)}$ is the unique global minimum. We introduce

$$U^{(1)} := z^*(U_2^{(0)}, U_1^{(0)}) .$$

According to Theorem 16.4 and its improvment (16.16), we obtain

$$\omega_1(V_\beta) = \frac{h^{-1+1/\mu}}{2\pi}|\widehat{\lambda}_1(U^{(1)})|\sqrt{\left|\frac{\det(\mathrm{Hess}\,V(U_2^{(0)}))}{\det(\mathrm{Hess}\,V(U^{(1)}))}\right|}$$

$$\times \exp -\frac{1}{h}\left(V(U^{(1)}) - V(U_2^{(0)})\right) \times (1 + \mathcal{O}(h)) \qquad (17.3)$$

$$= \frac{\beta^{1-1/\mu}}{2\pi}|\widehat{\lambda}_1(U^{(1)})|\sqrt{\left|\frac{\det(\mathrm{Hess}\,V(U_2^{(0)}))}{\det(\mathrm{Hess}\,V(U^{(1)}))}\right|}$$

$$\times \exp -\beta\left(V(U^{(1)}) - V(U_2^{(0)})\right) \times (1 + \mathcal{O}(\beta^{-1})) , \qquad (17.4)$$

where $\widehat{\lambda}_1(U^{(1)})$ denotes the unique negative eigenvalue of the Hessian of V at the saddle point $U^{(1)}$. Hence in the low temperature regime, that is when $\beta \ge \beta_V$, we get $\omega_1(V_\beta) \le 1$.

Theorem 17.1.
Assume that $V \in S(\langle x \rangle^{2\mu}, \frac{dx^2}{\langle x \rangle^2})$ and that $\langle \nabla V(x) \rangle$ is elliptic in the symbol class $S(\langle x \rangle^{2\mu - 1}, \frac{dx^2}{\langle x \rangle^2})$, $\mu > 1/2$. Assume additionally that it is a Morse function with two local minima and $\lim_{x \to \infty} V(x) = +\infty$. Then the rate of return to the equilibrium τ satisfies

$$\frac{\beta^{1-1/\mu}\rho(\beta)}{64(5 + 3\gamma_0\sqrt{m}\beta^{\frac{\mu-1}{2\mu}} + 3C_{V_\beta})^2} \leq \tau \leq c_V \beta^{3/2 - 1/2\mu}\sqrt{\rho(\beta)} \log\left[Q(\sqrt{m}\gamma_0, \beta, 1)\right] ,$$

for any $\beta \geq \beta_V$, where $c_V > 0$ and $\beta_V > 0$ only depend on V, the quantitity $\rho(\beta)$ is defined by

$$\omega_1(V_\beta) = \beta^{1-1/\mu}\rho(\beta) , \tag{17.5}$$

and C_{V_β} is defined by (17.2).

This result can be simplified if the potential is quadratic at infinity since in this case $\mu = 1$, hence $\beta^{1-1/\mu} = 1$, and

$$C_{V_\beta}^2 \leq \kappa_V(1 + \mathcal{O}_V(\beta^{-1/2}))$$

$$\text{with } \kappa_V := \max_{x_0, \nabla V(x_0)=0} \sigma\left[(\text{Hess } V(x_0))^2 + \frac{1}{2}|\Delta V(x_0)|\text{Id}\right] .$$

Corollary 17.2.
If V is quadratic at infinity the previous lower and upper bounds lead to

$$\frac{\rho(\beta)(1 + \mathcal{O}_V(\beta^{-1/2}))}{64(5 + 3\gamma_0\sqrt{m} + 3\sqrt{\kappa_V})^2} \leq \tau \leq c_V \beta \sqrt{\rho(\beta)} \log\left[Q(\sqrt{m}\gamma_0, \beta, 1)\right] .$$

18

Epilogue

Our aim in this text was not to give a definite treatment of the spectral and regularity properties of Fokker-Planck operators or Witten Laplacians. We tried instead to give an account of how the known techniques from partial differential equations and spectral theory can be applied for their analysis, while completing or referring to existing and sometimes recent results. We hope that this synthetic text will help the researchers in Partial Differential Equations, Probability theory or Mathematical Physics for further developments in this field, which happened to be and is still very active.

During the publishing process of this text some new results have been obtained. The accurate asymptotics of the exponentially small eigenvalues presented in Chapter 16 have been proved in a quite general framework in [HelKlNi] and [HelNi2]. An accurate description of the spectrum and pseudospectrum of a semiclassical Fokker-Planck operator has been given by Hérau-Sjöstrand-Stolk in [HerSjSt]. A work in preparation by F. Hérau deals with the return to the equilibrium for some nonlinear Fokker-Planck equation arising in kinetic theory.

When writing the final version of this text, we heared also about the recent work of Bismut [Bi2, Bi3, Bi4, Bi5] (and even more recently about his collaboration with Lebeau [Leb]). The so called "hypoelliptic Laplacian" that he introduces in order to compute geometrical invariants and which acts in the cotangent bundle (phase-space), looks like what we have called here the Fokker-Planck operator with a partial diffusion only in the momentum variable. The structures exhibited by Bismut bring a new point of view and may suggest new questions in analysis.

Besides the mathematical questions that we addressed in this text, other developments are possible towards more involved models: general drift-diffusion operators, chains of anharmonic oscillators, other kinetic equations. Our point of view was to restrict our attention to the simplest models which already exhibit a very rich structure. For further information on related problems or other issues, we refer the reader to [Ris], [EckHai1], [Re-BeTh3] or [Vi2].

References

[Ag] S. Agmon. *Lectures on exponential decay of solutions of second order elliptic equations.* Mathematical notes of Princeton university n^0 29 (1982).

[Aetall] C. Ané, S. Blachère, D. Chafaï, P. Fougères, I. Gentil, F. Malrieu, C. Roberto, and G. Scheffer. *Sur les inégalités de Sobolev logarithmiques (avec une préface de D. Bakry et M. Ledoux).* Panoramas & Synthèses, n^0 10 (2000). Société Mathématique de France.

[ArMaUn] A. Arnold, P.A. Markowich, and A. Unterreiter. On convex Sobolev inequalities and the rate of convergence to equilibrium for Fokker-Planck type equations. Comm. Partial Differential Equations, 26(1-2):43–100, (2001).

[Arn] V.I. Arnold. *Mathematical methods of classical mechanics.* Graduate Texts in Mathematics 60. Springer-Verlag, New York.

[ArGuVar] V.I. Arnold, S.M. Gusen-Zade, and A.N. Varchenko. *Singularities of Differentiable Maps. Vol II.* Monographs in Mathematics, Birkhaüser (1988).

[AvHeSi] J.E. Avron, I. Herbst, and B. Simon. Schrödinger operators with magnetic fields I. General interactions. Duke Math. J. 45 (4), p. 847-883 (1978).

[BaJeSj] V. Bach, T. Jecko, and J. Sjöstrand. Correlation asymptotics of classical lattice spin systems with nonconvex Hamilton function at low temperature. Ann. Henri Poincaré 1, p. 59-100 (2000).

[BaMo1] V. Bach and J. Schach Moeller. Correlation at low temperature I. Exponential decay. J. Funct. Anal. 203 (1), p. 93-148 (2003).

[BaMo2] V. Bach and J. Schach Moeller. Correlation at low temperature II. Asymptotics. To appear in J. Statist. Phys. 2004.

[Be1] R. Beals. A general calculus of pseudodifferential operators. Duke Math. J. 42 (1), p. 1-42 (1975).

[Be2] R. Beals : Characterization of p. d. o. and applications. Duke Math. J. 44 (1), p. 45-57 (1977), and Erratum Vol. 46 (1), Duke Math. J. (1979).

[Be3] R. Beals. Weighted distributions spaces and pseudodifferential operators. J. An. Math. 39, p. 130-187 (1981).

[BeSh] F.A. Berezin and M.A. Shubin. *The Schrödinger Equation.* Kluwer Academic Publishers, Mathematics and its Applications, Vol. 66, (1991).

[Bi1] J.M. Bismut. The Witten complex and the degenerate Morse inequalities. J. Differential Geom. 23 (3), p. 207-240 (1986).

[Bi2] J.M. Bismut. Une déformation de la théorie de Hodge sur le fibré cotangent. C.R. Acad. Sci. Paris Sér. I, 338, p. 471-476 (2004).

[Bi3] J.M. Bismut. Le Laplacien hypoelliptique sur le fibré cotangent. C.R. Acad. Sci. Paris Sér. I, 338, p. 555-559 (2004).

[Bi4] J.M. Bismut. Une déformation en famille du complexe de de Rham-Hodge. C.R. Acad. Sci. Paris Sér. I, 338, p. 623-627 (2004).

[Bi5] J.M. Bismut. The hypoelliptic laplacian on the cotangent bundle. preprint.

[BZ] J.M. Bismut and W. Zhang. *An extension of a theorem by Müller and Cheeger.* Astérisque 205 (1992). Société Mathématique de France.

[BodHel1] T. Bodineau and B. Helffer. LogSobolev inequality for unbounded spins systems. J. Funct. Anal. 166 (1), Aug. 1, p. 168-178 (1999).

[BodHel2] T. Bodineau and B. Helffer. Correlations, Spectral gap and LogSobolev inequalities for unbounded spins systems. Proceedings of the UAB Conference, March 16-20 (1999). Differential equations and mathematical physics, p. 51-66 (2000).

[BoBuRo] P. Boggiatto, E. Buzano, and L. Rodino. *Global hypoellipticity and spectral theory.* Mathematical Research, 92. Akademie Verlag, Berlin, (1996).

[BoCaNo] P. Bolley, J. Camus, and J. Nourrigat. La condition de Hörmander-Kohn pour les opérateurs pseudo-différentiels. Comm. Partial Differential Equations 7 (2), p. 197-221 (1982).

[BoDaHel] P. Bolley, M. Dauge, and B. Helffer. Conditions suffisantes pour l'injection compacte d'espaces de Sobolev à poids (ou autour d'une question de F. Mignot). Séminaire de l'université de Nantes 1989-90.

[Bon1] J.M. Bony. Unpublished graduate course 1997-1998. Book in preparation.

[Bon2] J.M. Bony. Caractérisations des opérateurs pseudo-différentiels. Séminaire sur les Equations aux Dérivées Partielles, 1996–1997, Exp. XXIII, Ecole Polytechnique, Palaiseau (1997).

[Bon3] J.M. Bony. Personal communication (Sept. 2003).

[BonChe] J.M. Bony and J.Y. Chemin. Espaces fonctionnels associés au calcul de Weyl-Hörmander. Bull. Soc. Math. France 122, n^0 1, p. 77-118 (1994).

[BoLe] J.M. Bony and N. Lerner. Quantification asymptotique et microlocalisation d'ordre supérieur I. Ann. Scient. Ec. Norm. Sup., 4e série 22, p. 377-433 (1989).

[Bou] L. S. Boulton. Non-self-adjoint harmonic oscillator, compact semigroups, and pseudospectra. J. Operator Theory 47, n^0 2, p. 413–429 (2002).

[BovEckGayKl1] A. Bovier, M. Eckhoff, V. Gayrard and M. Klein. Metastability in stochastic dynamics of disordered mean-field models. Prob. Theory and related fields 119 (1), p. 99-161 (2001).

[BovEckGayKl2] A. Bovier, M. Eckhoff, V. Gayrard and M. Klein. Metastability in reversible diffusion processes I. Sharp asymptotics for capacities and exit times. Journal of the European Mathematical Society (in Press) (2004).

[BovGayKl] A. Bovier, V. Gayrard and M. Klein. Metastability in reversible diffusion processes II Precise asymptotics for small eigenvalues. Preprint 2002 (new version in 2004), to appear in Journal of the European Mathematical Society (2004).

[Brin] H.C. Brinckman. Physica 22, 29 (1956).

[ChFu] M. Christ and S. Fu. Compactness in the $\bar{\partial}$-Neumann problem, magnetic Schrödinger operators, and the Aharonov-Bohm effect. Preprint (2003), math.CV/0311225.

[Cole] S. Coleman. The use of instantons. Proc. Int. School of Physics (Erice) (1977).

[CoDuSe] J.M. Combes, P. Duclos, and R. Seiler. Krein's formula and one dimensional multiple well. J. Funct. Anal. 52 (2), p. 257-301 (1983).

[CFKS] H.L Cycon, R.G Froese, W. Kirsch, and B. Simon. *Schrödinger operators with application to quantum mechanics and global geometry.* Text and Monographs in Physics. Springer-Verlag (1987).

[DaLi] R. Dautray and J.L. Lions. *Mathematical analysis and numerical methods for science and technology.* Springer Verlag (1992).

[Dav1] E.B. Davies. *One-parameter semigroups.* Academic Press Inc. [Harcourt Brace Jovanovich Publishers], London, (1980).

[Dav2] E.B. Davies. *Spectral theory and differential operators.* Cambridge Studies in Advanced Mathematics 42. Cambridge University Press, Cambridge (1995).

[Dav3] E.B. Davies. Pseudospectra, the harmonic oscillator and complex resonances. R. Soc. London Ser. A Math. Phys. Eng. Sci. 455, p. 585-599 (1999).

[Dav4] E.B. Davies. Semi-classical states for non self-adjoint Schrödinger operators. Comm. Math. Phys. 200, p. 35-41 (1999).

[Dav5] E.B. Davies. Non-self-adjoint differential operators. Bull. London Math. Soc. 34(5), p. 513-532 (2002).

[Den] N. Dencker. The Weyl calculus with locally temperate metrics and weights. Ark. Mat. 24 (1), p. 59-79 (1986).

[DeSjZw] N. Dencker, J. Sjöstrand, and M. Zworski. Pseudospectra of semi-classical (pseudo)differential operators. Comm. Pure Appl. Math. 57 (3), p. 384-415 (2004).

[Der] M. Derridj. Sur une classe d'opérateurs différentiels hypoelliptiques à coefficients analytiques. Séminaire Goulaouic-Schwartz 1970-1971: Equations aux dérivées partielles et analyse fonctionnelle, Exp. n° 12, Centre de Math., Ecole Polytech. (1971).

[DesVi] L. Desvillettes and C. Villani. On the trend to global equilibrium in spatially inhomogeneous entropy-dissipating systems : the linear Fokker-Planck equation. Comm. Pure Appl. Math. 54 (1), p. 1-42 (2001).

[DeuSt] J.D. Deuschel and D. Stroock. *Large Deviations.* Pure Appl. Math. 137, Boston, Academic Press (1989).

[DiSj] M. Dimassi and J. Sjöstrand. *Spectral Asymptotics in the semi-classical limit.* London Mathematical Society. Lecture Note Series 268. Cambridge University Press (1999).

[EckHai1] J.P. Eckmann and M. Hairer. Non-equilibrium statistical mechanics of strongly anharmonic chains of oscillators. Comm. Math. Physics 212 (1), p. 105-164 (2000).

[EckHai2] J.P. Eckmann and M. Hairer. Spectral properties of hypoelliptic operators. Comm. Math. Phys. 235 (2), p. 233-253 (2003).

[EckPiRe-Be] J.P. Eckmann, C.A. Pillet, and L. Rey-Bellet. Non-equilibrium statistical mechanics of anharmonic chains coupled to two heat baths at different temperatures. Comm. Math. Phys. 208 (2), p. 275-281 (1999).

[Fei] V.I. Feigin. New classes of peudo-differential operators on $\mathbb{R}^n$. Trudy Moskov. Mat. Obshch. 36, p. 155-194 (1978).

[Fol] G.B. Folland. On the Rothschild-Stein lifting theorem. Comm. Partial Differential Equations 2 (2), p. 165-191 (1977).

[FuSt] S. Fu and E. Straube. Semiclassical analysis and compactness in the $\bar{\partial}$-Neumann problem. J. Math. Ann. Appl. 271 (1), p. 267-282 (2002). Erratum in 280 (1), p. 195-196 (2003).

[Glo1] P. Glowacki. The Rockland condition for nondifferential convolution operators. Duke Math. J. 58 (2), p. 371-395 (1989).

[Glo2] P. Glowacki. The Rockland condition for nondifferential convolution operators. II. Studia Math. 98 (2), p. 99-114 (1991).

[Glo3] P. Glowacki. The Weyl asymptotic formula by the method of Tulovskiĭ and Shubin. Studia Math. 127 (2), p. 169-190 (1998).

[Goo] R. Goodman. *Nilpotent Lie groups: structure and applications to analysis.* Lecture Notes in Mathematics n° 562. Springer-Verlag, Berlin-New York, (1976).

[GHH] G.M. Graf, D. Hasler, and J. Hoppe. No zero energy states for the supersymmetric x^2y^2 potential. Lett. Math. Phys. 60, p. 191-196 (2002).

[Har] E. Harrell. On the rate of asymptotic eigenvalue degeneracy. Comm. Math. Phys. 60, p. 73-95 (1978).

[Hel0] B. Helffer. *Théorie spectrale pour des opérateurs globalement elliptiques.* Astérisque 112 (1984). Société Mathématique de France.

[Hel1] B. Helffer. Sur l'hypoellipticité des opérateurs de la forme $\sum Y_j^2 + \frac{1}{2}\sum c_{j,k}[Y_j, Y_k]$. Séminaire de l'université de Nantes, exposé n°1, 1981-82 (d'après Helffer-Métivier-Nourrigat).

[Hel2] B. Helffer. Partial differential equations on nilpotent groups. Lie group representations, III (College Park, Md., 1982/1983). Springer Lecture Notes in Mathematics n° 1077, p. 210-254 (1984).

[Hel3] B. Helffer. Conditions nécessaires d'hypoanalyticité pour des opérateurs invariants à gauche sur un groupe nilpotent gradué, Journal of differential Equations 44 (3), p. 460-481 (1982).

[Hel4] B. Helffer. *Introduction to the semi-classical Analysis for the Schrödinger operator and applications.* Springer Verlag. Lecture Notes in Math. n°1336 (1988).

[Hel5] B. Helffer. Etude du Laplacien de Witten associé à une fonction de Morse dégénérée. Publications de l'université de Nantes, Séminaire EDP 87-88.

[Hel6] B. Helffer. On spectral theory for Schrödinger operators with magnetic potentials. Advanced Studies in Pure Mathematics 23, p. 113-141 (1993).

[Hel7] B. Helffer. *Semiclassical analysis for Schrödinger operators, Laplace integrals and transfer operators in large dimension: an introduction.* Cours de DEA, *Paris Onze Edition* (1995).

[Hel8] B. Helffer. Recent results and open problems on Schrödinger operators, Laplace integrals and transfer operators in large dimension. Advances in PDE, Schrödinger operators, Markov semigroups, Wavelet analysis, Operator algebras, edited by M. Demuth, B.W. Schulze, E. Schrohe and J. Sjöstrand, Akademie Verlag, Vol 11, p. 11-162 (1996).

[Hel9] B. Helffer. Splitting in large dimension and infrared estimates. NATO ASI Series. Series C: Mathematical and Physical Sciences- Vol. 490. Kluwer Academic Publishers, p. 307-348 (1997).

[Hel10] B. Helffer. Splitting in large dimension and infrared estimates II Moment estimates. J. Math. Phys. 39 (2), p. 760-776 (1998).

[Hel11] B. Helffer. *Semi-classical analysis, Witten Laplacians and statistical mechanics.* World Scientific (2002).

[Hel12] B. Helffer. Estimations hypoelliptiques globales et compacité de la résolvante pour des opérateurs de Fokker-Planck ou des laplaciens de Witten. Actes du colloque franco-tunisien de Hammamet (Septembre 2003). Preprint mp_arc 04-57.

[HelKlNi] B. Helffer, M. Klein, and F. Nier. Quantitative analysis of metastability in reversible diffusion processes via a Witten complex approach. Preprint 03-25 IRMAR-Univ. Rennes 1, (2003).

[HelMo] B. Helffer and A. Mohamed. Sur le spectre essentiel des opérateurs de Schrödinger avec champ magnétique. Ann. Inst. Fourier 38(2), p. 95-113 (1988).

[HelNi1] B. Helffer and F. Nier. Criteria for the Poincaré inequality associated with Dirichlet forms in $\mathbb{R}^d$, $d \geq 2$. Int. Math. Res. Notices 22, p. 1199-1224 (2003).

[HelNi2] B. Helffer and F. Nier. Quantitative analysis of metastability in reversible diffusion processes via a Witten complex approach : the case with boundary. In preparation.

[HelNo1] B. Helffer and J. Nourrigat. Hypoellipticité pour des groupes nilpotents de rang 3. Comm. Partial Differential Equations 3 (8), p. 643-743 (1978).

[HelNo2] B. Helffer and J. Nourrigat. Caractérisation des opérateurs hypoelliptiques homogènes invariants à gauche sur un groupe nilpotent gradué. Comm. Partial Differential Equations 4 (8), p. 899-958 (1979).

[HelNo3] B. Helffer and J. Nourrigat. *Hypoellipticité maximale pour des opérateurs polynômes de champs de vecteur.* Progress in Mathematics, Birkhäuser, Vol. 58 (1985).

[HelNo4] B. Helffer and J. Nourrigat. Remarques sur le principe d'incertitude. J. Funct. Anal. 80 (1), p. 33-46 (1988).

[HeNoWa] B. Helffer, J. Nourrigat, and X.P. Wang. Spectre essentiel pour l'équation de Dirac. Annales Scientifiques de l'Ecole Normale Supérieure 4ème série, t. 22, p. 515-533 (1989).

[HelSj1] B. Helffer and J. Sjöstrand. Puits multiples en limite semi-classique. Comm. Partial Differential Equations 9 (4), p. 337-408 (1984).

[HelSj2] B. Helffer and J. Sjöstrand. Puits multiples en limite semi-classique II -Interaction moléculaire-Symétries-Perturbations. Annales de l'IHP (section Physique théorique) 42 (2), p. 127-212 (1985).

[HelSj3] B. Helffer and J. Sjöstrand. Multiple wells in the semi-classical limit III. Math. Nachrichten 124, p. 263-313 (1985).

[HelSj4] B. Helffer and J. Sjöstrand. Puits multiples en limite semi-classique IV -Etude du complexe de Witten -. Comm. Partial Differential Equations 10 (3), p. 245-340 (1985).

[HelSj5] B. Helffer and J. Sjöstrand. A proof of the Bott inequalities. In Volume in honor of M. Sato, Algebraic Analysis, Vol.1, p. 171-183. Academic Press (1988).

[HelSj6] B. Helffer and J. Sjöstrand. Analyse semiclassique pour l'équation de Harper. Bull. SMF, 116(4), mémoire 34 (1988).

[HelSj7] B. Helffer and J. Sjöstrand. Semiclassical expansions of the thermodynamic limit for a Schrödinger equation. Actes du colloque Méthodes semi-classiques à l'université de Nantes, 24 Juin-30 Juin 1991, Astérisque 210, p. 135-181 (1992).

[HelSj8] B. Helffer and J. Sjöstrand. On the correlation for Kac like models in the convex case. J. Statist. Phys. 74 (1-2), p. 349-369 (1994).

[HerNi] F. Hérau and F. Nier. Isotropic hypoellipticity and trend to the equilibrium for the Fokker-Planck equation with high degree potential. Archive for Rational Mechanics and Analysis 171 (2), p. 151-218 (2004).

[HerSjSt] F. Hérau, J. Sjöstrand and C. Stolk. Semiclassical analysis for the Kramers-Fokker-Planck equation. preprint.

[HolKusStr] R. Holley, S. Kusuoka, and D. Stroock. Asymptotics of the spectral gap with applications to the theory of simulated annealing. J. Funct. Anal. 83 (2), p. 333-347 (1989).

[Hor1] L. Hörmander. Hypoelliptic second order differential equations. Acta Mathematica 119, p. 147-171 (1967).

200   References

[Hor2] L. Hörmander. *The analysis of linear partial differential operators.* Springer Verlag (1985).

[Iwas] K. Iwasaki. Isolated singularity, Witten's Laplacian and duality for twisted De Rham cohomology. Comm. Partial Differential Equations 28, n° 1&2, p. 61-82 (2003).

[Iw] A. Iwatsuka. Magnetic Schrödinger operators with compact resolvent. J. Math. Kyoto Univ. 26, p. 357-374 (1986).

[Jo] J. Johnsen. On the spectral properties of Witten Laplacians, their range projections and Brascamp-Lieb's inequality. Integral Equations Operator Theory 36 (3), p. 288-324 (2000).

[JoMaSc81] G. Jona-Lasinio, F. Martinelli, and E. Scoppola : New approach to the semiclassical limit of quantum mechanics. I. Multiple tunnelings in one dimension. Comm. Math. Phys. 80 (2), p. 223-254 (1981).

[Kac] M. Kac. Mathematical mechanisms of phase transitions. Brandeis lectures, Gordon and Breach (1966).

[Ki] A. Kirillov. Unitary representations of nilpotent groups. Russian Math. Surveys 17, p. 53-104 (1962) .

[Ko] J. Kohn. Lectures on degenerate elliptic problems. Pseudodifferential operators with applications, C.I.M.E., Bressanone 1977, p. 89-151 (1978).

[Kol] A.N. Kolmogorov. Zufällige Bewegungen. Ann. of Math. (2) 35, p. 116-117 (1934).

[KonShu] V. Kondratiev and M. Shubin. Discreteness of spectrum for the magnetic Schrödinger operators. Comm. Partial Differential Equations 27 (3-4), p. 477-526 (2002).

[Lau] F. Laudenbach. *Topologie Différentielle.* Cours de Majeure de l'Ecole Polytechnique (1995).

[Leb] G. Lebeau. Analyse du Laplacien hypoelliptique. Séminaire Equations aux Dérivées Partiellles, Ecole Polytechnique (oct. 2004).

[Ler] N. Lerner: The Wick calculus of pseudo-differential operators and energy estimates. New trends in microlocal analysis (Tokyo, 1995), p. 23-37, Springer, Tokyo (1997).

[Mai1] H.M. Maire. Hypoelliptic overdetermined systems of partial differential equations. Comm. Partial Differential Equations 5 (4), p. 331-380 (1980).

[Mai2] H.M. Maire. Résolubilité et hypoellipticité de systèmes surdéterminés. Séminaire Goulaouic-Schwartz 1979-1980, Exp. V, Ecole Polytechnique (1980).

[Mai3] H.M. Maire. Necessary and sufficient condition for maximal hypoellipticity of $\bar{\partial}_b$. Unpublished (1979).

[Mai4] H.M. Maire. Régularité optimale des solutions de systèmes différentiels et du Laplacien associé: application au $\square_b$. Math. Ann. 258, p. 55-63 (1981).

[Mar] A. Martinez. *An introduction to semiclassical and microlocal analysis.* Universitext. Springer-Verlag, New York, (2002).

[Mas84] V.P. Maslov. Global exponential asymptotics of the solutions of the tunnel equations and the large deviation problems (Russ.). Tr. Mosk. Inst. Akad. Nauk 163, p.150-180 (1984). (Engl. transl. in Proc. Stecklov Inst. Math 4 (1985).)

[MaMo] O. Matte and J.S. Moeller. On the spectrum of semi-classical Witten-Laplacians and Schrödinger operators in large dimension. Preprint mp_arc 03 448 (2003).

[Mef] M. Meftah. Conditions suffisantes pour la compacité de la résolvante d'un opérateur de Schrödinger avec un champ magnétique. J. Math. Kyoto Univ. 31 (3), p. 875-880 (1991).

[Mel] A. Melin. Parametrix constructions for right invariant differential operators on nilpotent groups. Ann. Global Anal. Geom. 1 (1), p. 79-130 (1983).

[Mi] L. Miclo. Comportement de spectres d'opérateurs à basse température. Bull. Sci. Math. 119, p. 529-533 (1995).

[Mil] J. Milnor. *Morse theory*. Princeton University Press (1963).

[Moh] A. Mohamed. Comportement asymptotique, avec estimation du reste, des valeurs propres d'une classe d'opérateurs pseudo-différentiels sur $\mathbb{R}^n$. Math. Nachr. 140, p. 127-186 (1989).

[MoNo] A. Mohamed and J. Nourrigat. Encadrement du $N(\lambda)$ pour des opérateurs de Schrödinger avec champ magnétique. J. Math. Pures Appl. (9) 70, n^0 1, p. 87-99 (1991).

[NaNi] F. Nataf and F. Nier. Convergence of domain decomposition methods via semi-classical calculus. Comm. Partial Differential Equations 23 (5-6), p. 1007-1059 (1998).

[Ni1] F. Nier. Quelques critères pour l'inégalité de Poincaré dans $\mathbb{R}^d$, $d \geq 2$. Séminaire Equations aux Dérivées Partiellles, 2002-2003, exposé V, Ecole Polytechnique (2003).

[Ni2] F. Nier. A variational formulation of Schrödinger-Poisson systems in dimension $d \leq 3$. Comm. Partial Differential Equations 18 (7-8), p. 1125-1147 (1993).

[Ni3] F. Nier. Quantitative analysis of metastability in reversible diffusion processes via a Witten complex approach. Conférence Equations aux Dérivées Partielles in Forges-Les-Eaux, (2004).

[No1] J. Nourrigat. *Subelliptic estimates for systems of pseudo-differential operators*. Course in Recife (1982). University of Recife.

[No2] J. Nourrigat. Inégalités L^2 et représentations de groupes nilpotents. J. Funct. Anal. 74 (2), p. 300-327 (1987).

[No3] J. Nourrigat. Réduction microlocale des systèmes d'opérateurs pseudo-différentiels. Ann. Inst. Fourier 36 (3), p. 83-108 (1986).

[No4] J. Nourrigat. Systèmes sous-elliptiques. Séminaire Equations aux Dérivées Partielles, 1986-1987, exposé V, Ecole Polytechnique (1987).

[No5] J. Nourrigat. L^2 inequalities and representations of nilpotent groups. CIMPA school of harmonic analysis, Wuhan (1991).

[No6] J. Nourrigat. Subelliptic systems. Comm. Partial Differential Equations 15 (3), p. 341-405 (1990).

[No7] J. Nourrigat. Subelliptic systems II. Invent. Math. 104 (2), p. 377-400 (1991).

[ReSi] M. Reed and B. Simon. *Method of Modern Mathematical Physics*. Academic press, (1975).

[Re-BeTh1] L. Rey-Bellet and L.E. Thomas. Asymptotic behavior of thermal nonequilibrium steady states for a driven chain of anharmonic oscillators. Comm. Math. Phys. 215 (1), p. 1-24 (2000).

[Re-BeTh2] L. Rey-Bellet and L.E. Thomas. Exponential convergence to nonequilibrium stationary states in classical statistical mechanics. Comm. Math. Phys. 225 (2), p. 305-329 (2000).

[Re-BeTh3] L. Rey-Bellet and L.E. Thomas. Fluctuations of the entropy production in anharmonic chains. Ann. Henri Poincaré 3 (3), p. 483-502 (2002).

[RiNa] F. Riesz and B. Sz.-Nagy. *Functional analysis*. Dover Publications Inc., New York, (1990).

[Ris] H. Risken. *The Fokker-Planck equation. Methods of solution and applications*. Springer-Verlag, Berlin, second edition (1989).

[Rob] D. Robert. Propriétés spectrales d'opérateurs pseudodifférentiels. Comm. Partial Differential Equations 3 (9), p. 755-826 (1978).

[Rob2] D. Robert. Comportement asymptotique des valeurs propres d'opérateurs du type Schrödinger à potentiel "dégénéré". J. Math. Pures Appl. (9) 61, n°. 3, p. 275–300 (1982).

[Roc] C. Rockland. Hypoellipticity on the Heisenberg group, representation theoretic criteria. Trans. Amer. Math. Soc. 240, p. 1-52 (1978).

[RoSt] L.P. Rothschild and E.M. Stein. Hypoelliptic differential operators and nilpotent groups. Acta Mathematica 137, p. 248-315 (1977).

[See] R. Seeley. Complex powers of an elliptic operator and singular integrals. Proc. Symposia Pure Math. 10. AMS, p. 288-307 (1967).

[She] Zhongwei Shen. Eigenvalue asymptotics and exponential decay of eigenfunctions for Schrödinger operators with magnetic fields. Trans. Amer. Math. Soc. 348 (11), p. 4465-4488 (1996).

[Shu] M. Shubin. *Pseudodifferential operators and spectral theory*. Translated from the 1978 Russian original. Second edition. Springer-Verlag, Berlin (2001).

[Sima] C.G. Simader. Essential self-adjointness of Schrödinger operators bounded from below. Math. Z. 159, p. 47-50 (1978).

[Sim1] B. Simon. Some quantum operators with discrete spectrum but classically continuous spectrum, Ann. Physics 146, p. 209-220 (1983).

[Sim2] B. Simon. Semi-classical analysis of low lying eigenvalues, I. Nondegenerate minima : Asymptotic expansions. Ann. Inst. Poincaré (Section Physique Théorique) 38, p. 296-307 (1983).

[Sim3] B. Simon. *Trace ideals and their applications*. Cambridge University Press IX, Lecture Notes Series vol. 35 (1979).

[Sj1] J. Sjöstrand. *Singularités analytiques microlocales*. Astérisque 95 (1982), Société Mathématique de France.

[Sj2] J. Sjöstrand. Potential wells in high dimensions I. Ann. Inst. Poincaré (Section Physique théorique) 58 (1), p. 1-41 (1993).

[Sj3] J. Sjöstrand. Potential wells in high dimensions II, more about the one well case. Ann. Inst. Poincaré (Section Physique théorique) 58 (1), p. 42-53 (1993).

[Sj4] J. Sjöstrand. Correlation asymptotics and Witten Laplacians, St Petersburg Math. J. 8, p. 160-191 (1997).

[Sj5] J. Sjöstrand. *Complete asymptotics for correlations of Laplace integrals in the semi-classical limit*. Mém. Soc. Math. France 83 (2000).

[Sj6] J. Sjöstrand. Pseudospectrum for differential operators. Séminaire Equations aux Dérivées Partielles 2002-2003, Exposé XVI, Ecole Polytechnique (2003).

[Sm1] S. Smale. Morse inequalities for a dynamical system. Bull. Amer. Math. Soc. 66, p. 373-375 (1960).

[Sm2] S. Smale. On gradient dynamical systems. Ann. of Math. 74, p. 199-206 (1961).

[Ta1] D. Talay. Approximation of invariant measures of nonlinear Hamiltonian and dissipative stochastic differential equations. In C. Soize R. Bouc, editor, *Progress in Stochastic Structural Dynamics*, volume 152 of *Publication du L.M.A.-C.N.R.S.*, p 139-169 (1999).

[Ta2] D. Talay. Stochastic Hamiltonian dissipative systems with non globally Lipschitz coefficients: exponential convergence to the invariant measure and discretization by the implicit Euler scheme. Markov Process. Related Fields 8(2), p. 163-198 (2002).

[Tha] B. Thaller. *The Dirac equation.* Texts and monographs in Physics. Springer
Verlag (1992).

[Tref] L.N. Trefethen. Pseudospectra of linear operators. Siam Review 39 (3),
p. 383-406 (1997).

[Tr1] F. Trèves. An invariant criterion of hypoellipticity. Amer. J. Math., Vol. 83,
p. 645-668 (1961).

[Tr2] F. Trèves. Study of a model in the theory of complexes of pseudo-differential
operators. Ann. of Maths (2) 104, p. 269-324 (1976). See also erratum: Ann.
Math. (2) 113, p. 423 (1981).

[TuSh] V.N. Tulovskii and M. Shubin. The asymptotic behavior of the eigenvalues
of pseudodifferential operators in L^2 ($\mathbb{R}^n$). Uspehi Mat. Nauk 28, n^0 5 (173),
p. 242 (1973).

[Vi1] C. Villani. Régularité hypoelliptique et convergence vers l'équilibre pour des
équations cinétiques non linéaires. Lecture at the Workshop CinHypWit in
Rennes. Feb. 2003.

[Vi2] C. Villani. A review of mathematical topics in collisional kinetic theory. *Hand-
book of Fluid Mechanics*, Vol. 1. S. Friedlander and D. Serre (eds.). North-
Holland (2002).

[Wi] E. Witten. Supersymmetry and Morse inequalities. J. Diff. Geom. 17, p. 661-
692 (1982).

[Yos] K. Yosida. *Functional Analysis*, Grundlehren der mathematischen Wis-
senschaften in Einzeldarstellungen, Vol. 123. Springer-Verlag (1968).

[ZH] W. Zhang. *Lectures on Chern-Weil theory and Witten deformations.* Nankai
Tracts in Mathematics, Vol. 4, World Scientific (2001).

[Zwo] M. Zworski. Numerical linear algebra and solvability of partial differential
equations. Comm. Math. Phys. 229 (2), p. 293-307 (2002).

Index

Printing and Binding: Strauss GmbH, Mörlenbach

Lecture Notes in Mathematics

For information about Vols. 1–1673
please contact your bookseller or Springer

Vol. 1780: J. H. Bruinier, Borcherds Products on O(2,1) and Chern Classes of Heegner Divisors (2002)

Vol. 1781: E. Bolthausen, E. Perkins, A. van der Vaart, Lectures on Probability Theory and Statistics. Ecole d' Eté de Probabilités de Saint-Flour XXIX-1999. Editor: P. Bernard (2002)

Vol. 1782: C.-H. Chu, A. T.-M. Lau, Harmonic Functions on Groups and Fourier Algebras (2002)

Vol. 1783: L. Grüne, Asymptotic Behavior of Dynamical and Control Systems under Perturbation and Discretization (2002)

Vol. 1784: L.H. Eliasson, S. B. Kuksin, S. Marmi, J.-C. Yoccoz, Dynamical Systems and Small Divisors. Cetraro, Italy 1998. Editors: S. Marmi, J.-C. Yoccoz (2002)

Vol. 1785: J. Arias de Reyna, Pointwise Convergence of Fourier Series (2002)

Vol. 1786: S. D. Cutkosky, Monomialization of Morphisms from 3-Folds to Surfaces (2002)

Vol. 1787: S. Caenepeel, G. Militaru, S. Zhu, Frobenius and Separable Functors for Generalized Module Categories and Nonlinear Equations (2002)

Vol. 1788: A. Vasil'ev, Moduli of Families of Curves for Conformal and Quasiconformal Mappings (2002)

Vol. 1789: Y. Sommerhäuser, Yetter-Drinfel'd Hopf algebras over groups of prime order (2002)

Vol. 1790: X. Zhan, Matrix Inequalities (2002)

Vol. 1791: M. Knebusch, D. Zhang, Manis Valuations and Prüfer Extensions I: A new Chapter in Commutative Algebra (2002)

Vol. 1792: D. D. Ang, R. Gorenflo, V. K. Le, D. D. Trong, Moment Theory and Some Inverse Problems in Potential Theory and Heat Conduction (2002)

Vol. 1793: J. Cortés Monforte, Geometric, Control and Numerical Aspects of Nonholonomic Systems (2002)

Vol. 1794: N. Pytheas Fogg, Substitution in Dynamics, Arithmetics and Combinatorics. Editors: V. Berthé, S. Ferenczi, C. Mauduit, A. Siegel (2002)

Vol. 1795: H. Li, Filtered-Graded Transfer in Using Noncommutative Gröbner Bases (2002)

Vol. 1796: J.M. Melenk, hp-Finite Element Methods for Singular Perturbations (2002)

Vol. 1797: B. Schmidt, Characters and Cyclotomic Fields in Finite Geometry (2002)

Vol. 1798: W.M. Oliva, Geometric Mechanics (2002)

Vol. 1799: H. Pajot, Analytic Capacity, Rectifiability, Menger Curvature and the Cauchy Integral (2002)

Vol. 1800: O. Gabber, L. Ramero, Almost Ring Theory (2003)

Vol. 1801: J. Azéma, M. Émery, M. Ledoux, M. Yor (Eds.), Séminaire de Probabilités XXXVI (2003)

Vol. 1802: V. Capasso, E. Merzbach, B.G. Ivanoff, M. Dozzi, R. Dalang, T. Mountford, Topics in Spatial Stochastic Processes. Martina Franca, Italy 2001. Editor: E. Merzbach (2003)

Vol. 1803: G. Dolzmann, Variational Methods for Crystalline Microstructure - Analysis and Computation (2003)

Vol. 1804: I. Cherednik, Ya. Markov, R. Howe, G. Lusztig, Iwahori-Hecke Algebras and their Representation Theory. Martina Franca, Italy 1999. Editors: V. Baldoni, D. Barbasch (2003)

Vol. 1805: F. Cao, Geometric Curve Evolution and Image Processing (2003)

Vol. 1806: H. Broer, I. Hoveijn. G. Lunther, G. Vegter, Bifurcations in Hamiltonian Systems. Computing Singularities by Gröbner Bases (2003)

Vol. 1807: V. D. Milman, G. Schechtman (Eds.), Geometric Aspects of Functional Analysis. Israel Seminar 2000-2002 (2003)

Vol. 1808: W. Schindler, Measures with Symmetry Properties (2003)

Vol. 1809: O. Steinbach, Stability Estimates for Hybrid Coupled Domain Decomposition Methods (2003)

Vol. 1810: J. Wengenroth, Derived Functors in Functional Analysis (2003)

Vol. 1811: J. Stevens, Deformations of Singularities (2003)

Vol. 1812: L. Ambrosio, K. Deckelnick, G. Dziuk, M. Mimura, V. A. Solonnikov, H. M. Soner, Mathematical Aspects of Evolving Interfaces. Madeira, Funchal, Portugal 2000. Editors: P. Colli, J. F. Rodrigues (2003)

Vol. 1813: L. Ambrosio, L. A. Caffarelli, Y. Brenier, G. Buttazzo, C. Villani, Optimal Transportation and its Applications. Martina Franca, Italy 2001. Editors: L. A. Caffarelli, S. Salsa (2003)

Vol. 1814: P. Bank, F. Baudoin, H. Föllmer, L.C.G. Rogers, M. Soner, N. Touzi, Paris-Princeton Lectures on Mathematical Finance 2002 (2003)

Vol. 1815: A. M. Vershik (Ed.), Asymptotic Combinatorics with Applications to Mathematical Physics. St. Petersburg, Russia 2001 (2003)

Vol. 1816: S. Albeverio, W. Schachermayer, M. Talagrand, Lectures on Probability Theory and Statistics. Ecole d'Eté de Probabilités de Saint-Flour XXX-2000. Editor: P. Bernard (2003)

Vol. 1817: E. Koelink, W. Van Assche(Eds.), Orthogonal Polynomials and Special Functions. Leuven 2002 (2003)

Vol. 1818: M. Bildhauer, Convex Variational Problems with Linear, nearly Linear and/or Anisotropic Growth Conditions (2003)

Vol. 1819: D. Masser, Yu. V. Nesterenko, H. P. Schlickewei, W. M. Schmidt, M. Waldschmidt, Diophantine Approximation. Cetraro, Italy 2000. Editors: F. Amoroso, U. Zannier (2003)

Vol. 1820: F. Hiai, H. Kosaki, Means of Hilbert Space Operators (2003)

Vol. 1821: S. Teufel, Adiabatic Perturbation Theory in Quantum Dynamics (2003)

Vol. 1822: S.-N. Chow, R. Conti, R. Johnson, J. Mallet-Paret, R. Nussbaum, Dynamical Systems. Cetraro, Italy 2000. Editors: J. W. Macki, P. Zecca (2003)

Vol. 1823: A. M. Anile, W. Allegretto, C. Ringhofer, Mathematical Problems in Semiconductor Physics. Cetraro, Italy 1998. Editor: A. M. Anile (2003)

Vol. 1824: J. A. Navarro González, J. B. Sancho de Salas, C^∞ - Differentiable Spaces (2003)

Vol. 1825: J. H. Bramble, A. Cohen, W. Dahmen, Multiscale Problems and Methods in Numerical Simulations, Martina Franca, Italy 2001. Editor: C. Canuto (2003)

Vol. 1826: K. Dohmen, Improved Bonferroni Inequalities via Abstract Tubes. Inequalities and Identities of Inclusion-Exclusion Type. VIII, 113 p, 2003.

Vol. 1827: K. M. Pilgrim, Combinations of Complex Dynamical Systems. IX, 118 p, 2003.

Vol. 1828: D. J. Green, Gröbner Bases and the Computation of Group Cohomology. XII, 138 p, 2003.

Recent Reprints and New Editions

4. Manuscripts should in general be submitted in English. Final manuscripts should contain at least 100 pages of mathematical text and should always include
   - a general table of contents;
   - an informative introduction, with adequate motivation and perhaps some historical remarks: it should be accessible to a reader not intimately familiar with the topic treated;
   - a global subject index: as a rule this is genuinely helpful for the reader.

5. Lecture Notes volumes are, as a rule, printed digitally from the authors' files. We strongly recommend that all contributions in a volume be written in the same LaTeX version, preferably LaTeX2e. To ensure best results, authors are asked to use the LaTeX2e style files available from Springer's web-pages at

   www.springeronline.com

   [on this page, click on <Mathematics>, then on <For Authors> and look for <Macro Packages for books>]. Macros in LaTeX2.09 and TeX are available on request from: lnm@springer.de. Careful preparation of the manuscripts will help keep production time short besides ensuring satisfactory appearance of the finished book in print and online. After acceptance of the manuscript authors will be asked to prepare the final LaTeX source files (and also the corresponding dvi-, pdf- or zipped ps-file) together with the final printout made from these files. The LaTeX source files are essential for producing the full-text online version of the book. For the existing online volumes of LNM see: http://www.springerlink.com .
   The actual production of a Lecture Notes volume takes approximately 8 weeks.

6. Volume editors receive a total of 50 free copies of their volume to be shared with the authors, but no royalties. They and the authors are entitled to a discount of 33.3 % on the price of Springer books purchased for their personal use, if ordering directly from Springer.

7. Commitment to publish is made by letter of intent rather than by signing a formal contract. Springer-Verlag secures the copyright for each volume. Authors are free to reuse material contained in their LNM volumes in later publications: A brief written (or e-mail) request for formal permission is sufficient.

Addresses:

Professor J.-M. Morel, CMLA,
École Normale Supérieure de Cachan,
61 Avenue du Président Wilson, 94235 Cachan Cedex, France
E-mail: Jean-Michel.Morel@cmla.ens-cachan.fr

Professor F. Takens, Mathematisch Instituut,
Rijksuniversiteit Groningen, Postbus 800,
9700 AV Groningen, The Netherlands
E-mail: F.Takens@math.rug.nl

Professor B. Teissier, Université Paris 7
Institut Mathématique de Jussieu, UMR 7586 du CNRS
Équipe "Géométrie et Dynamique", 175 rue du Chevaleret
75013 Paris, France
E-mail: teissier@math.jussieu.fr

Springer-Verlag, Mathematics Editorial, Tiergartenstr. 17,
69121 Heidelberg, Germany,
Tel.: +49 (6221) 487-8410
Fax: +49 (6221) 487-8355
E-mail: lnm@springer.de